200 Challenging Math Problems

every 5th grader should know

This book belongs to:

....

Grade: ....

5

Guaranteed to improve children's math and success at school.

200 Challenging Math Problems

every 5th grader should know

New edition 2012

email: contactus@learn-2-think.com

ISBN: 978-981-07-2766-6

Master Grade 5 Math Problems

Introduction:

Solving math problems is core to understanding math concepts. When Math problems are presented as real-life problems students get a chance to apply their Math knowledge and skills. Word problems progressively develop a student's ability to visualize and logically interpret Mathematical situations.

This book provides numerous opportunities to every student to practice their math skills and develop their confidence of being a lifelong problem solver. The multi-step problem solving exercises in the book involve several math concepts. Student will learn more from these problems solving exercises than doing ten worksheets on the same math concepts.

The book is divided into 9 chapters. Within each chapter questions move from simple to advance word problems pertaining to the topic. The last chapter of the book explains step wise solutions to all the problems to reinforce learning and understanding.

How to use the book:

Here is a suggested plan that will help you to crack every problem in this book and outside.

Follow these 4 steps and all the Math problems will be a NO PROBLEM!

Read the problem carefully:

- What do I need to find out?
- What math operation is needed to solve the problem? For example addition, subtraction, multiplication, division etc.
- What clues and information do I have?
- What are the key words like sum, difference, product, perimeter, area, etc.?
- Which is the non-essential information?

Decide a plan

- Develop a plan based on the information that you have to solve the problem. Consider various strategies of problem solving:
- Drawing a model or picture
- Making a list
- Looking for pattern
- Working backwards
- Guessing and checking
- Using logical reasoning

Solve the problem:

Carry out the plan using the Math operation or formula you choose to find the answer.

Check your answer

- Check if the answer looks reasonable
- Work the problem again with the answer
- Remember the units of measure with the answer such as feet, inches, meter etc.

Master Grade 5 Math Problems

Note to the Teachers and Parents:

- ✎ Help students become great problem solvers by modelling a systematic approach to solve problems. Display the 'Four step plan of problem solving' for students to refer to while working independently or in groups.
- ✎ Emphasise on some key points in the problem.
- ✎ Enable students to enjoy the process of problem solving rather than being too focused on finding the answers.
- ✎ Provide opportunities to the students to think; explain and interpret the problem.
- ✎ Lead the student or the group to come up with the right strategy to solve the problem.
- ✎ Discuss the importance of showing steps of their work and checking their answers.
- ✎ Explore more than one possible solution to the problems.
- ✎ Give a chance to the students to present their work.

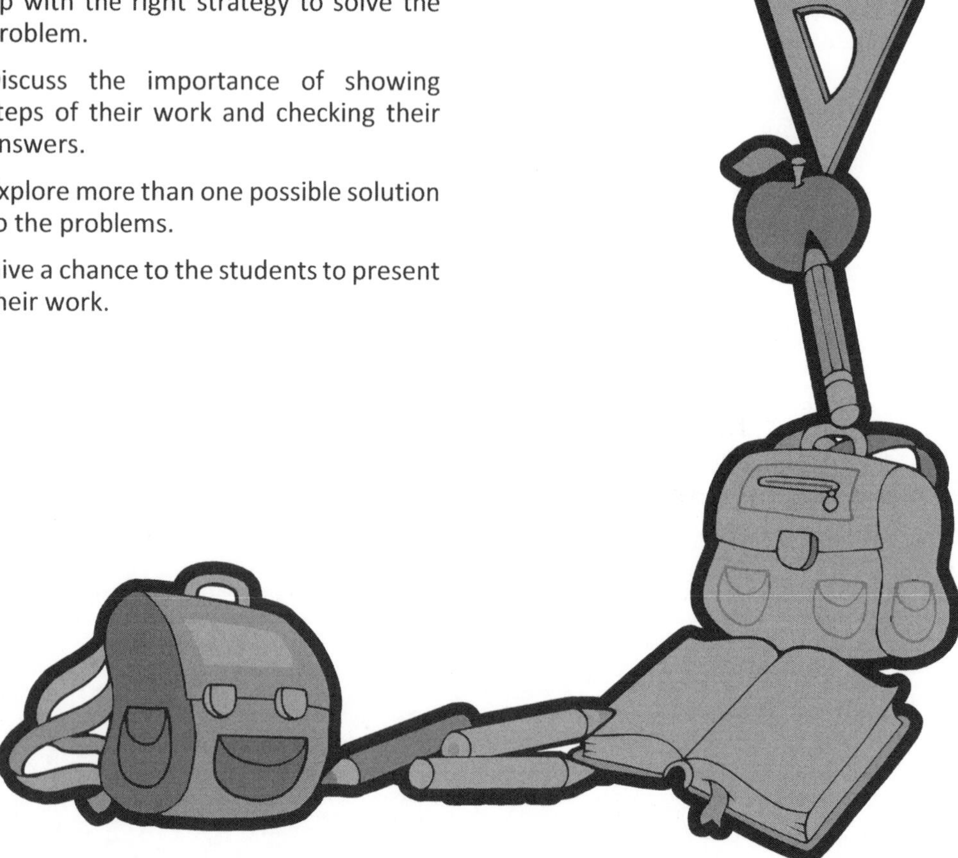

Contents

Topics **Page number**

NUMBERS

5

Guaranteed to improve children's Math and success at school.

There are 15 prime numbers less than 50.
How many prime numbers are less than 70?

Answer: ....

What percentage of the numbers from 1 to 100 are prime?

Answer: ....

A palindrome is any work or number which reads the same forwards or backwards. For example, the number "1881" and the word "level" are both palindromes. How many palindrome numbers can you find between 100 and 1000?

Answer: ....

How many times does the digit 2 appear in the prime numbers less than 100?

Answer: ....

What is the sum of the greatest 3-digit number and the greatest 4-digit number?

What is the difference between the greatest 4-digit number and the greatest 3-digit number?

Answer: ....

A positive number is called a perfect square if it is the square of a whole number. The first three perfect squares are 1, 4, and 9. What is the 100th perfect square?

Answer: ....

I am a number between 190 and 207. When divided by 4, the remainder is 2 and when divided by 5, the remainder is 1. What number am I?

Answer: ....

How many whole numbers less than 100 can be divided by 5 or 7?

Answer: ....

If you make all the four-digit numbers using each of the digits 4, 5, 8, and 9 exactly once; how many numbers will be a multiple of 4?

Answer: ....

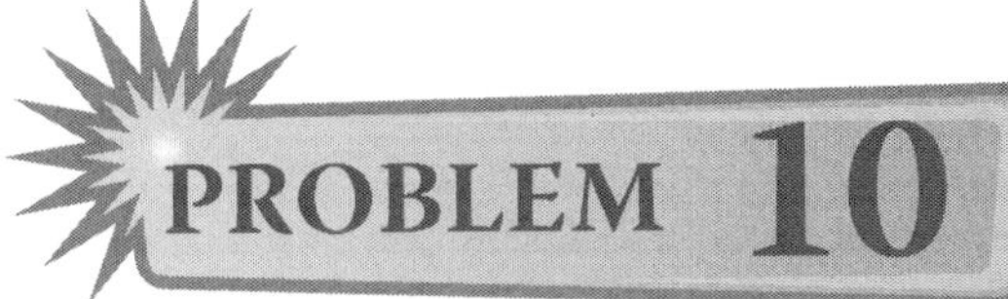

I am a number between 190 and 207. When divided by 4, the remainder is 2 and when divided by 5, the remainder is 1. What number am I?

Answer: ....

Which secret number is this? It is a three digit number. Its digits are in descending order. It is divisible by three. The hundred's digit is two more than the one's (unit's) digit. Adding the three digits will give you a dozen.

Answer: ....

2435X9 is a number that is divisible by 9. The digits in the tens place is covered and marked with an X.

What is the value of X?

Answer: ....

How many of the integers from 2 to 999 are squares of a whole number?

Answer: ....

The average of 7 consecutive numbers is 2009. What is the average of the first three of these numbers?

Answer: ....

Dividing 11 and 21 by the same number 'X' results in the same remainder.

What is the largest possible number that 'X' could be?

Answer: ....

A number when divided by 76 gives a quotient of 205 and a remainder of 13. What is the number?

Answer: ....

If you add the remainders of 77235 ÷ 7, 11111 ÷3 and 100001 ÷2 and divide that result by 7, what is the remainder?

Answer: ....

Two numbers are written on a white board. 1/6 of the first number is 8 more than 1/2 of the second number. The sum of the two numbers is 120.

a) What is the first number?

b) What is the second number?

Answer: ....

Place any of the four operation signs (+, -, x, ÷) between the nine digits so that the answer is equal to 100.

1 2 3 4 5 6 7 8 9 = 100

Answer: ....

Find four consecutive odd numbers which add to 88.

Answer: ....

The sum of two numbers is 15. The difference between the same two numbers is 3. What are the numbers?

Answer: ....

When two numbers are added, the sum is 27. One of the numbers is half of the other. What are the two numbers?

Answer: ....

A boat cannot carry more than 150 kg. Four friends weigh 60 kg, 80 kg, 80 kg, and 80 kg and they have to cross a river using the boat. What is the least number of trips necessary to carry the four friends across the river?

Answer: ....

Tracy recorded the amount of time in minutes it took to drive from her house to her office during a five-day period. Her driving times in minutes are shown below.

65, 57, 56, 59, 66

What is the mean (average) number of minutes it took Tracy to drive from her house to her office?

Answer: ....

Sarah got either a 90 or a 100 on each of her 5 math tests. The average of all her math tests is 98. How many 90s did she get?

Answer: ....

Kevin has 20 small balls of different colors: yellow, green, blue and black.

17 of the balls are not green, 5 are black, 12 are not yellow. How many blue balls does Kevin have?

Answer: ....

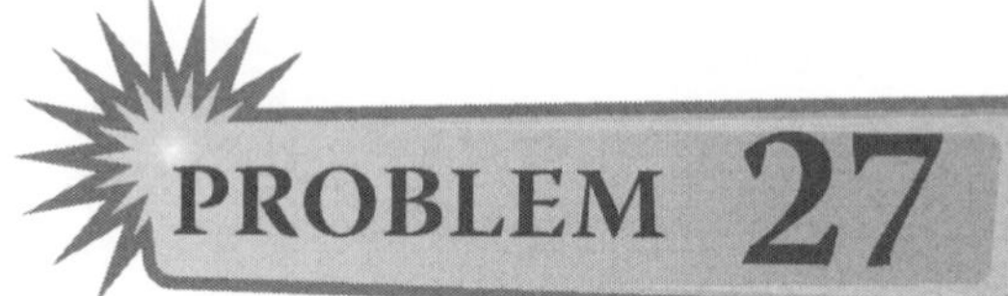

A Street is 350 meters long. Palm trees are planted on both sides of the street from the beginning to the end of the street at 5 meters apart. How many palm trees are planted?

Answer: ....

A snail is at the bottom of a 6 meter well. Each day the snail walks up 2.5 meters but at night it slips down 1.5 meters. On which day will the snail get out of the well?

Answer: ....

The following diagram shows the first three patterns of squares in a sequence. How many squares are there in the 50th pattern?

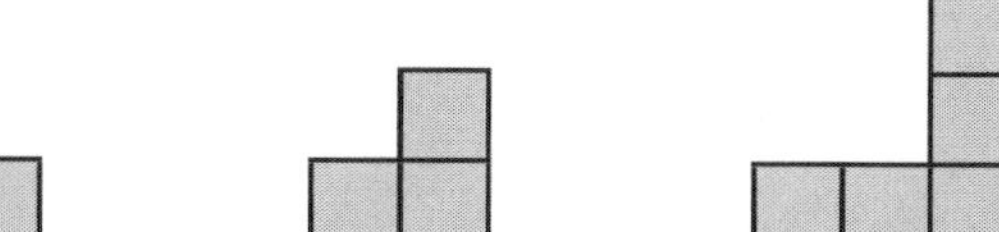

Answer: ....

There are 19 trees along the road from Sandra's home to school. Sandra marked some trees with a red strip as follows. On her way to school she marked the first tree, and then every second tree. On her way back, she marked the first tree, and then every third tree. How many trees have no mark on them?

Answer: ....

Twenty lamp posts are placed in a straight line. The distance between any two consecutive posts is 6 meters. What is the distance between the 1st and the 10th lamp post?

Answer: ....

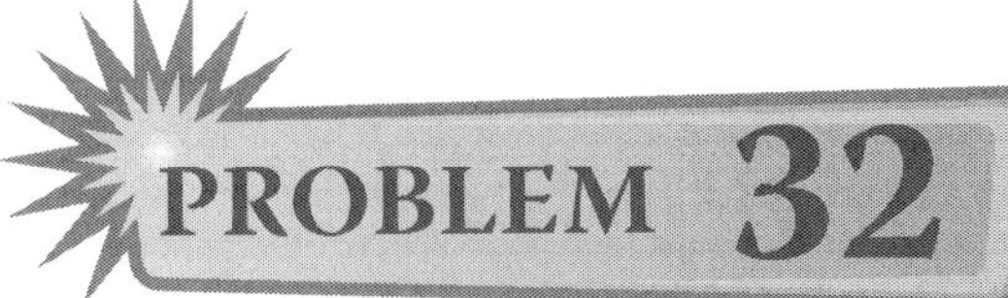

Points A, B, and C lie on a straight line, and A is not between B and C. The distance from A to B is 15 cm. The distance from C to A is 8cm. What is the distance from B to C?

Answer: ....

3 girls have an average height of 150 centimeters. 4 boys have an average height of 170 centimeters. What is the average height of all the girls and boys put together?

Answer: ....

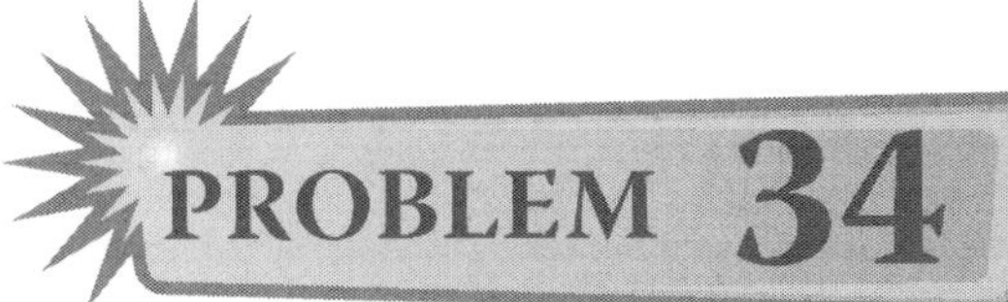

Sandra took the elevator to the 8th floor, then down 5 floors, up 10 floors, and finally down 1 floor. If those moves left Sandra on the middle floor of the building, how many floors are in the building?

Answer: ....

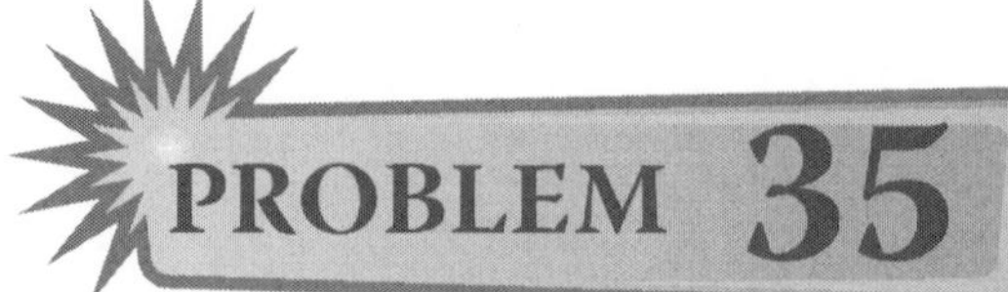

Suzy goes for her music lessons every 9 days, Amie goes for her music lessons every 12 days, and Bernard every 15 days. What is the least number of days before all three will have their music lessons on the same day?

Answer: ....

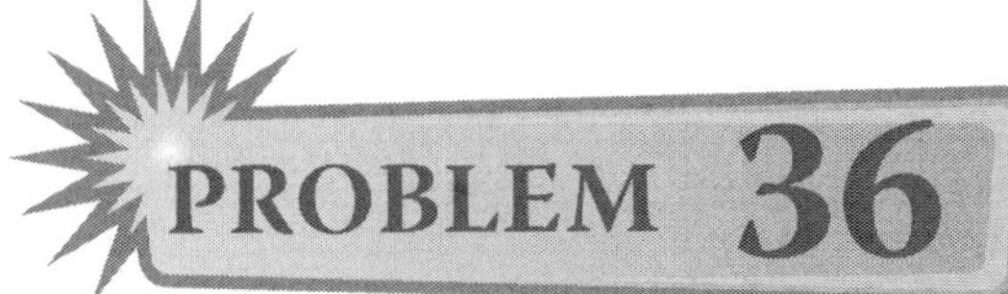

The average temperature from Monday to Wednesday was 34°C and the average temperature from Thursday to Saturday was 37°C. If the average of all 7 days, Monday to Sunday was 35°C, then what was the temperature on Sunday?

Answer: ....

In three math exams, Tim scores 84, 92, and 94. What score will Tim need to get in the fourth exam in order to have an average score of 91 for all the four exams?

Answer: ....

A box of pencils can be divided equally among 4, 5, 6 or 7 children with no pencils left. What is the least number of pencils that the box could have?

Answer: ....

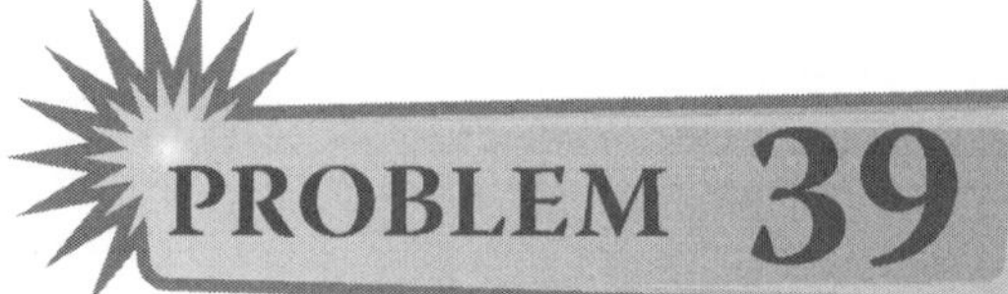

George has fewer than 500 books. When he divided his books into groups of three, there was one extra book left. When he divided his books into groups of four, there was still one extra book. He then divided his books into groups of five, but ended with the same result. He tried dividing his books into groups of seven, but still there was one extra book. How many books does George have?

Answer: ....

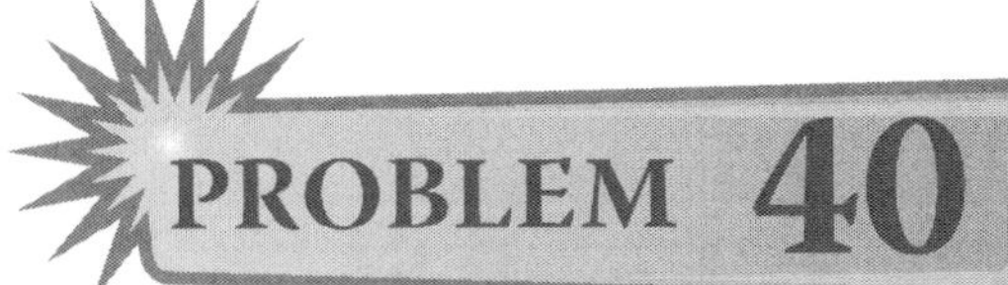

If I divide some chocolates among 15 children, each child gets 20 chocolates. If I divide the same number of chocolates among 10 children, how many chocolates will each child get?

Answer: ....

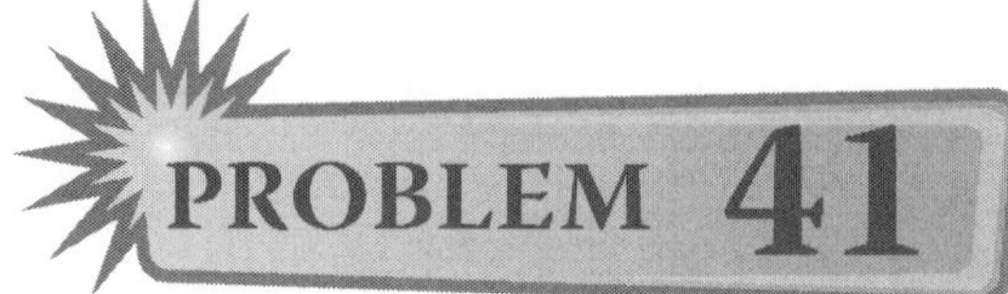

John is 15 cms shorter than James. James is 11 cms taller than Kate who is 14 cms shorter than Rebecca. Rebecca was 128 cms last year and in the last year she has grown by 7 cms. How tall is John?

Answer: ....

PROBLEM 42

The bells of church A ring every 6 minutes. The bells of church B ring every 9 minutes. The bells of church C ring every 15 minutes and the bells of church D ring every 20 minutes. If the bells of all the 4 churches rang at the same time at 6:00 A.M., at what time will the bells ring again at the same time?

Answer: ....

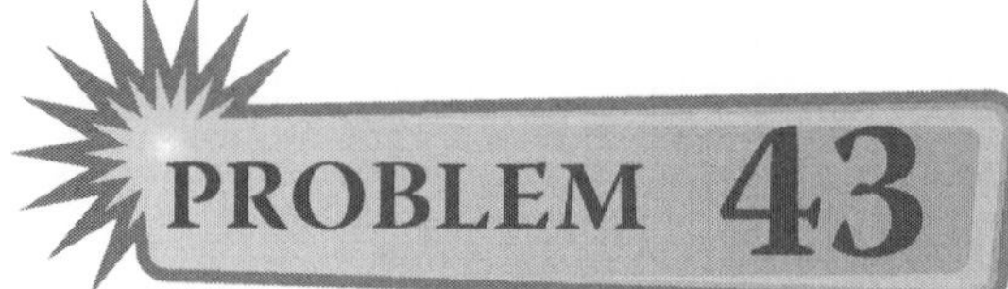

Felicia, Vivian, Shawn and Ella were born on March 1st, May 17th, July 20th and March 20th but not in that order. Vivian and Shawn were born in the same month, and Felicia's and Shawn's birthdays fall on the same dates in different months. Who was born on May 17th?

Answer: ....

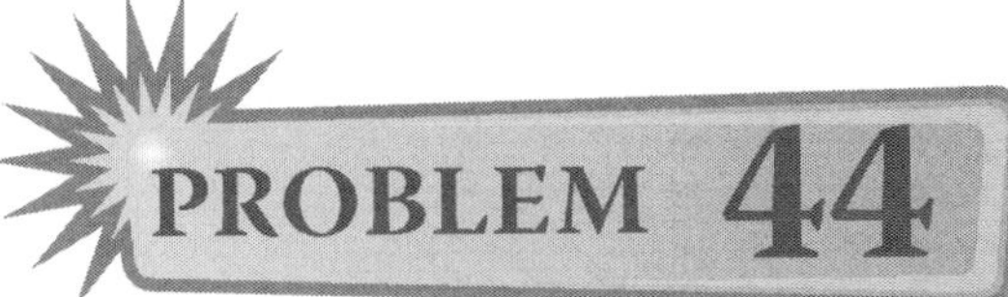

In a class of 30 students, exactly 7 have tape recorder, exactly 15 have pocket calculator, and exactly 2 have both. How many of the 30 students have neither?

Answer: ....

PROBLEM 45

The weight of a box of apples is 50 kg when it is full. When half of the apples in the box are taken out, the weight of the remaining apples and the box is 35 kg. What is the weight of the empty box?

Answer: ....

Rachel scored a total of 356 marks for Math, Science, English and Social Studies. She scored 97 marks for Science and 86 marks for English.

a) Find her total marks for Social Studies and Math

b) She scored 25 more marks for Math than for Social Studies, find her Math score

Answer: ....

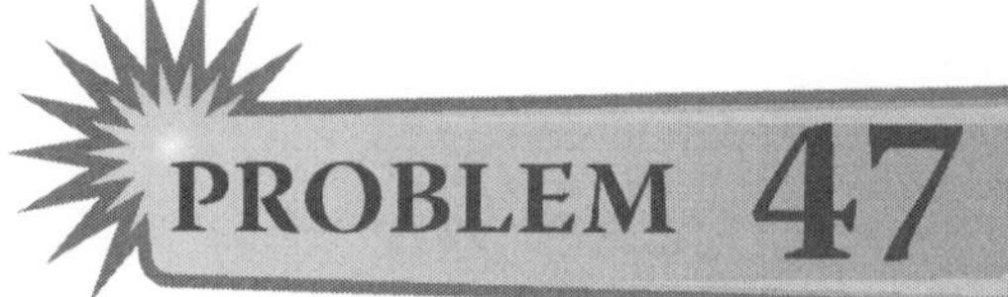

A total of 25 students participated in the school's sports day. 15 students participated in the school's orchestra. Six students participated in both. How many students participated in at least one event? How many only took up orchestra? How many took up only sports?

Answer: ....

Mike opens a book and sees the two page numbers on the facing pages. The product of these numbers is 420. What is the number of the page on the left side?

Answer: ....

ALGEBRA

5

Guaranteed to improve children's math and success at school.

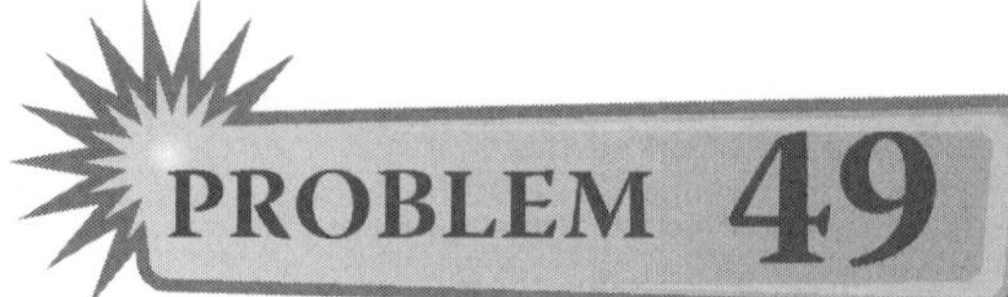

I threw 27 coins into the air. If twice as many coins landed heads up than tails up, how many coins landed heads up?

Answer: ....

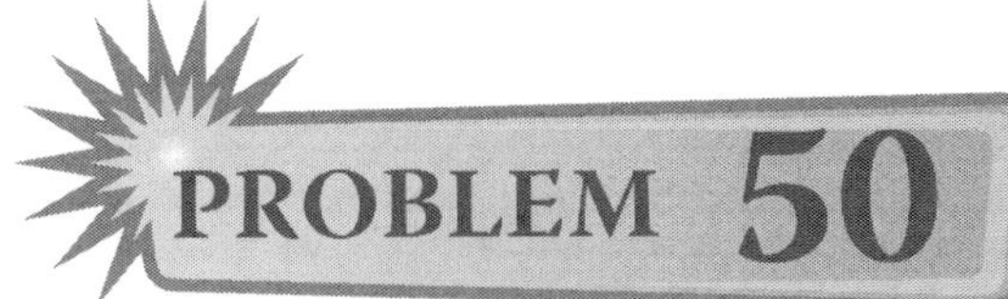

The only animals on farmer Jim's farm are cows and chicken. If Jim counts a total of 55 heads and a total of 136 feet, find the number of cows on Jim's farm.

Answer: ....

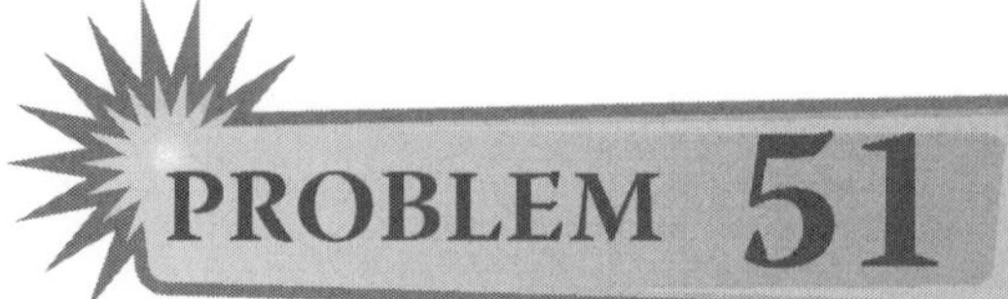

Jennifer is buying juice cans for her son's birthday party. Juice cans are sold by the carton. If she buys 3 cartons of juice cans, she will be short of 3 cans. If she buys 4 cartons, she will have 3 cans extra. How many cans are in each carton?

Answer: ....

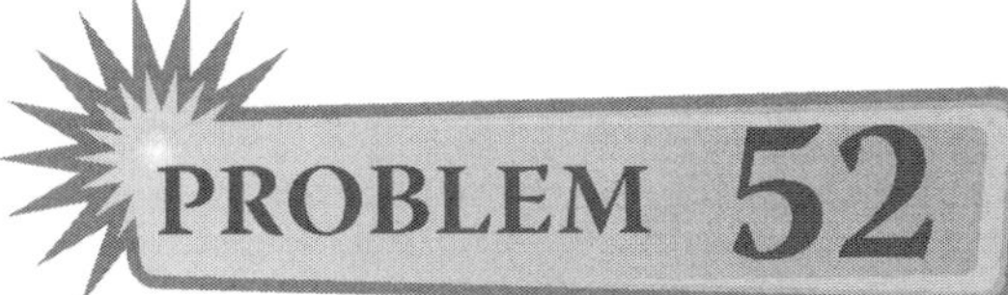

There are 24 students in Mrs Smith's class. Mandarin and French are the only two foreign language classes available, and everyone has to take at least one foreign language class. Nine students are taking Mandarin and 19 are taking French. How many of Mrs Smith's students are taking both Mandarin and French?

Answer: ....

There are twice as many boys in a room as girls. If 5 boys leave the room there would be an equal number of boys and girls in the room. How many boys were in the room at first?

Answer: ....

Henry had twice as many marbles as Jason. Kenneth had 3 times as many marbles as Jason. They had 846 marbles altogether. How many marbles did Kenneth have?

Answer: ....

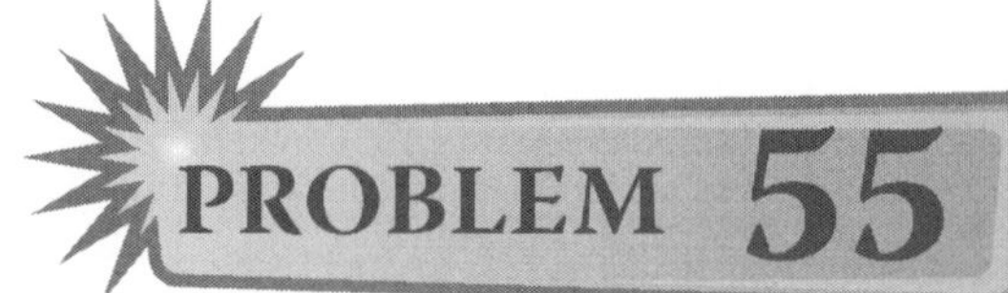

Jessica has some stamps. When she arranges them into 5 albums with 15 stamps in each album, she has 25 stamps left. She then arranges them into different albums with 5 stamps and 15 stamps in each album. How many albums does Jessica need if she wants to have same number of 5 stamps and 15 stamp albums?

Answer: ....

Alison bought 50 potatoes. She used 2/5 of them to bake. She used some potatoes to bake curry puffs. She had 12 potatoes left. How many potatoes did she use for curry puffs?

Answer: ....

Alice and Susan had an equal number of balloons. After Alice used 16 balloons, Susan had 3 times as many balloons as Alice. How many balloons did Susan have in the beginning?

Answer: ....

The total weight of 3 big balls and 4 small balls is 80 kg. Each big ball weighs twice as much as a small ball. Find the weight of a big ball.

Answer: ....

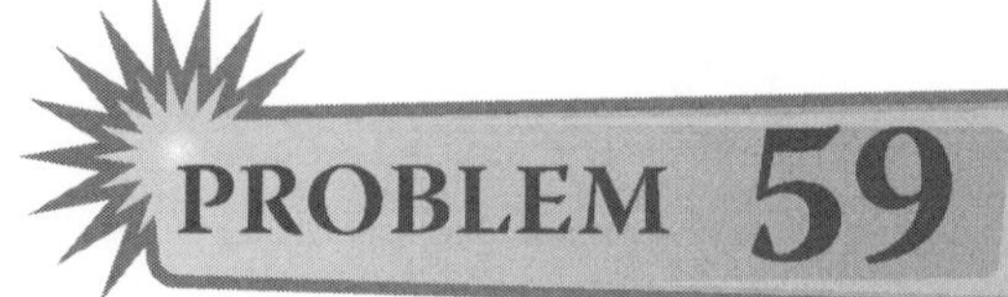

Ben had 150 stamps and Samuel had 50 stamps. After each of them bought an equal number of stamps, Ben had twice as many stamps as Samuel. How many stamps did each of them buy?

Answer: ....

One large bottle and two small bottles can hold the same amount of water as 6 small bottles. If one small bottle holds 1.5 liters water, then how much water does one large bottle hold?

Answer: ....

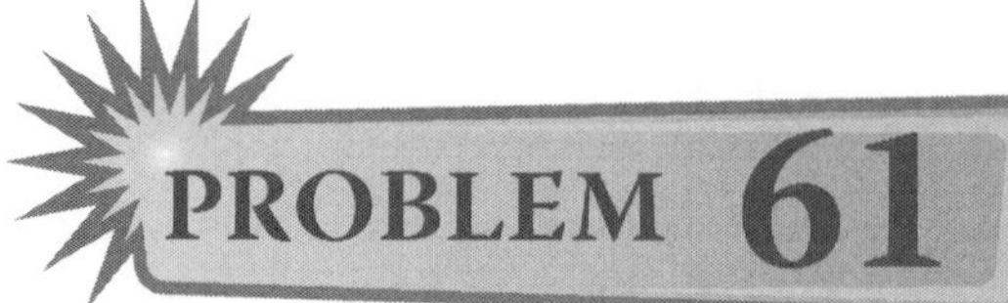

Sarah bought a candy jar from a shop. In the jar there were 6 more green candies than red, and one-third as many red candies as yellow candies. Sarah ate up all the 18 yellow candies first. How many green candies were still in the jar?

Answer: ....

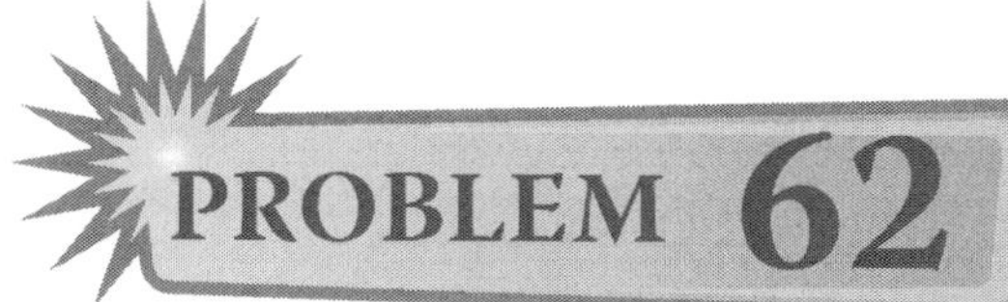

Jennifer has 89 pencils and erasers in her stationary drawer. She has 33 fewer erasers than pencils in her collection. How many pencils and erasers does she have?

Answer: ....

There were 5 times as many strawberries as blue berries in a container. Brandon ate half the number of strawberries and 20 strawberries were left in the container. How many strawberries and blueberries were in the container at first?

Answer: ....

Jessie had twice as many coins as Lisa. Jessie had 20 fewer coins than James. They had 50 coins altogether. How many coins did Lisa have?

Answer: ....

William spent 4 times as much on a magazine than on a towel. He had twice as much left as the amount of money he spent. He had $90 left. How much did the magazine cost?

Answer: ....

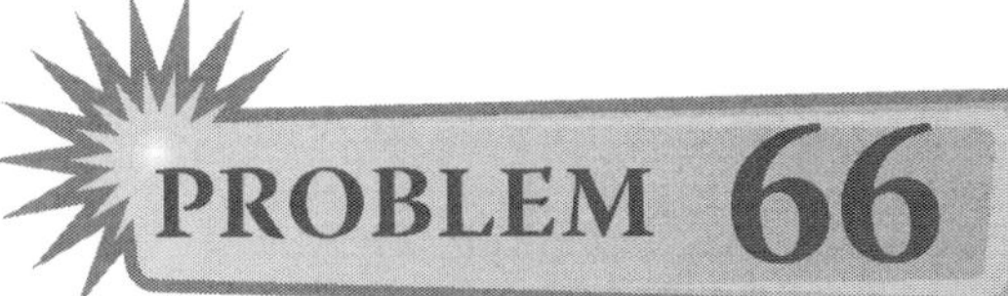

There were 720 orange cakes and twice as many yellow cakes. The rest were green cakes. If there were 2500 cakes altogether, how many green cakes were there?

Answer: ....

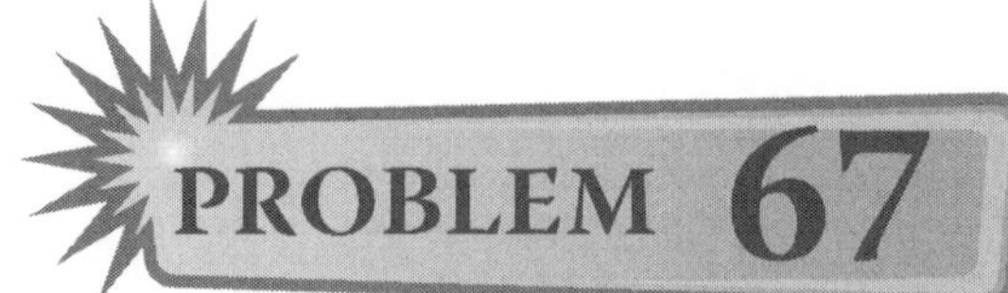

Janice spent half of her salary on food and transport and gave the remaining amount to her brother and savings in a bank. She gave $450 to her brother and saved $380. What was Janice's salary?

Answer: ....

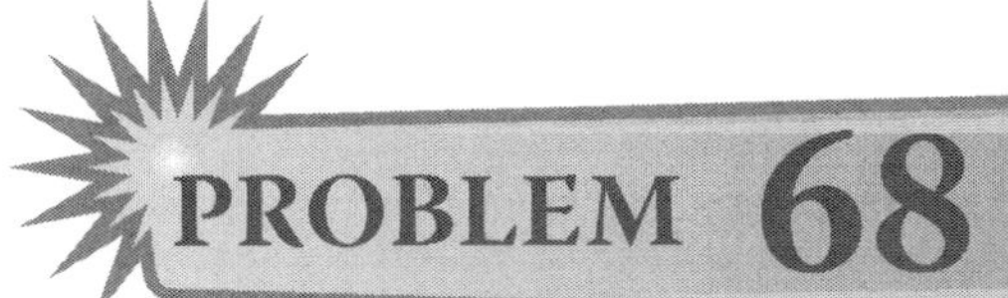

A total of 3560 people were on a trip to Malaysia. 1870 of them were adults, and the rest were children. If there were 490 more boys than girls, how many girls were there?

Answer: ....

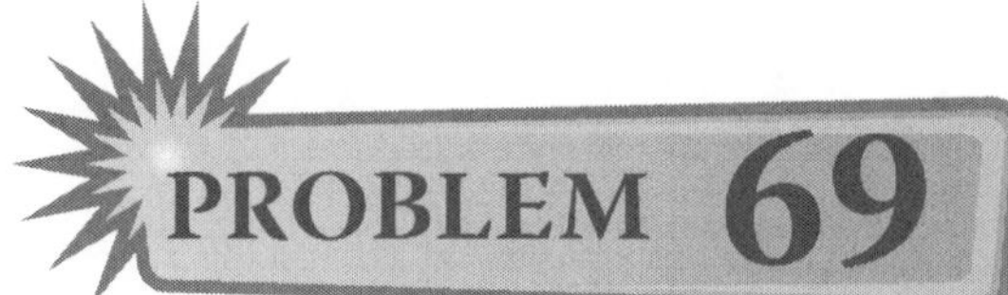

Anna gave 5 pencils to each of her friends. She then had 7 pencils left. If Anna had bought 8 packets of 4 pencils each, how many friends did she give the pencils to?

Answer: ....

At a party, there were 7 groups of 35 children. Another 4 boys and 16 girls joined the party. How many children were there altogether?

Answer: ....

FRACTIONS , DECIMAL AND PERCENTAGES

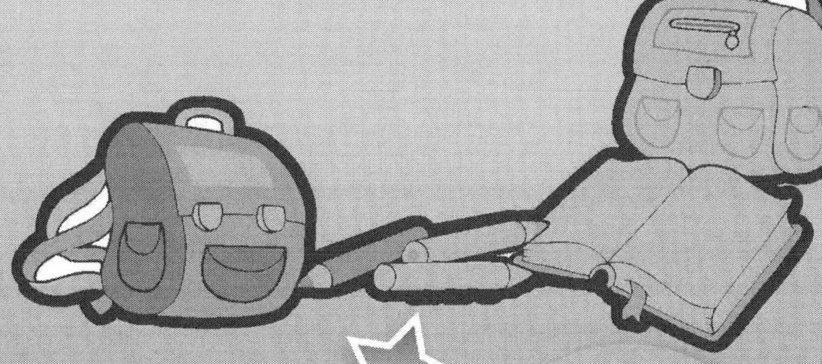

5

Guaranteed to improve children's math and success at school.

A school drama club has 150 members. 60% of the members are boys and the rest are girls. How many more boys than girls are there in the club?

Answer: ....

Gavin gave half of his sweets to Tim. Tim gave half of his sweets to Emma. Emma gave 1/4 of the sweets to Lily and kept the remaining 12. How many sweets did Gavin have in the beginning?

Answer: ....

In order to choose the class prefect for grade 5 at least 25% of the total students must be present to vote. There are 18 students in Grade 5. How many students must be present to elect the class prefect?

Answer: ....

A movie theatre is full. 46 girls occupy 2/5 of the seats. What is the capacity of the movie theatre?

Answer: ....

Tim plays tennis quarter of the time that Ben plays. Ben plays only two-fifth of the time that Katie plays. Katie plays tennis twice as long as Jake. If Jake plays Tennis for 5 hours, how long does Tim play?

Answer: ....

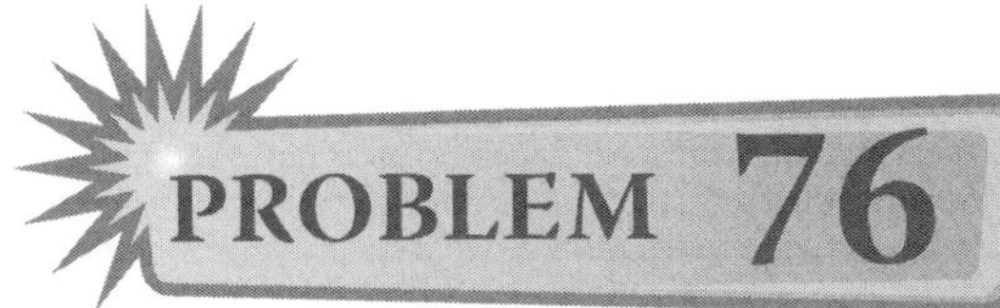

Christian added 3 grams sugar to 15 liters of water. What is the percentage of sugar in the solution obtained?

Answer: ....

Daniel has 200 books in his house library. 70 of them are fiction books.

What % of Daniel's books are non-fiction books?

Answer: ....

A 10 by 10 grid has 42% squares shaded.
How many squares are not shaded?

Answer: ....

In the election for class president, Lily received 40% of the votes, Felicia received 35% of the votes, and Claire received the rest of the votes. What percent did Claire receive?

Answer: ....

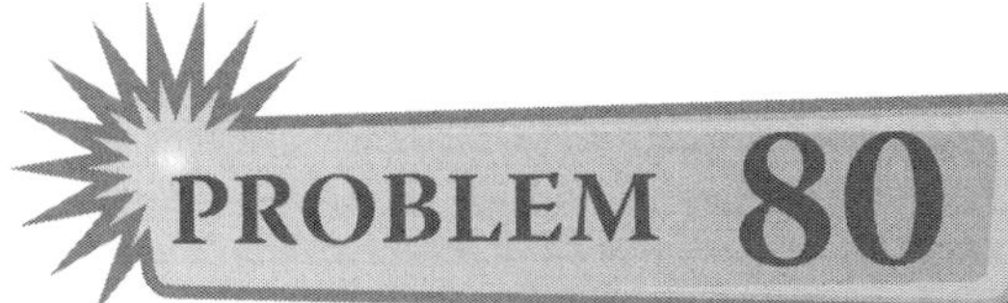

In a survey of children's favorite ice cream, chocolate chip received 22% of the votes. Strawberry received ¼ of the votes and vanilla received 0.37 of the votes. Order the flavors from most to least popular.

Answer: ....

Mrs. Franklin baked a total of 1500 apple tarts. She gave 45% of them to her relatives, sold 60% of the remaining tarts and gave 90 tarts to her neighbor. How many tarts did she have left?

Answer: ....

Some students were surveyed on their favorite subject. Two fifth students preferred math, 0.45 preferred English, and 48% preferred science. Order the subjects from least favorite to most favorite.

Answer: ....

Julian got the scores of his final exam. The average of his science, english, and history scores is 65%. When his math, science, english, and history grades are combined, the average score is 66%. What is his math score?

Answer: ....

PROBLEM 84

A school bus goes around dropping children at their bus stops. At the first stop, 1/4 of the students got off. At the second stop, 2/3 of the remaining students got off. At the third stop, 1/2 of the remaining students got off and at the fourth stop, all the remaining students got off. What is the least number of students that could have been on the bus?

Answer: ....

At 8:30 A.M., 200 students started running for a charity run. By 9:30 A.M., the number of students increased by 30%. At 10:30 A.M., 15% more students joined. After another hour, 25% more students joined. How many students in all ran for the charity run?

Answer: ....

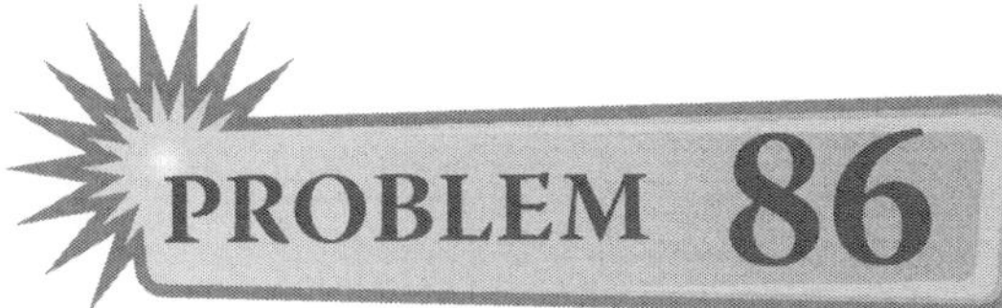

Mr. Richard spent 1/5 of his salary on food and 2/3 on clothes. Half of the remaining amount was given to his wife and the rest was saved. If he saved $520, what was Mr. Richard's salary?

Answer: ....

PROBLEM 87

Samuel sold 1/4 of his eggs on Monday and 1/4 of the remainder on Tuesday. The rest were sold on Wednesday. If he sold 144 eggs on Wednesday, how many eggs did he have in the beginning?

Answer: ....

Anna read 3/4 of a storybook. After reading, she discovered she still had 135 pages left to read. How many pages had she read?

Answer: ....

A rope is 3 meters long. It is divided into 9 parts. What length of the whole rope are the 2 parts of the rope?

Answer: ....

$\frac{2}{3}$ of a number is greater than $\frac{1}{2}$ of the same number by 12. What is the difference between the number and its 3rd multiple?

Answer: ....

There were a total of 2250 seats in a theatre. 10% of the seats were first class, 30% of the seats were second class and the rest of the seats were third class. 2 years later, 100 first class seats and 125 second class seats were added.

a) How many first class seats were there in the end?

b) How many second class seats were there in the end?

c) What percentage of seats were third class in the end?

Answer: ....

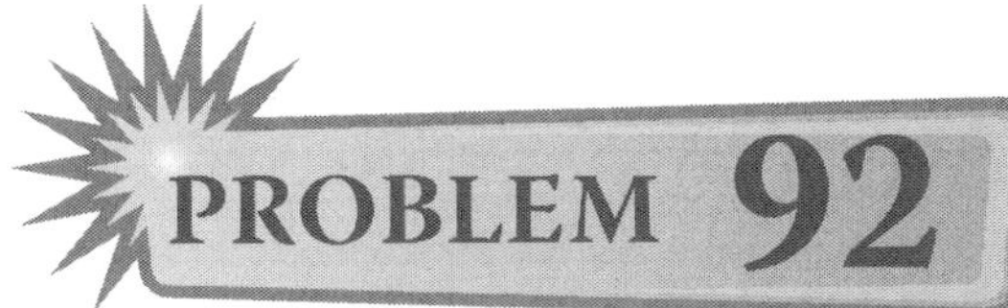

Mrs. Smith spent 2/3 of her savings to buy furniture for her house. She then spent 1/2 of her remaining savings to get it painted. If the cost of painting the house was $350, what were her original savings?

Answer: ....

RATIO AND PROPORTION

5

Guaranteed to improve children's math and success at school.

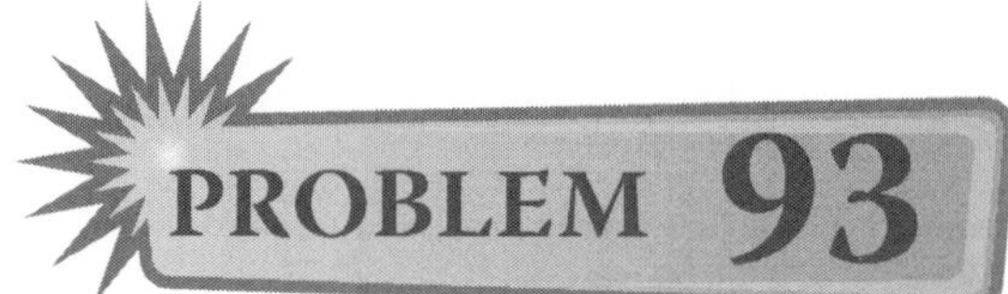

A coffee machine makes 150 cups of coffee every 7 minutes. How many minutes does it take this machine to make 525 cups?

Answer: ....

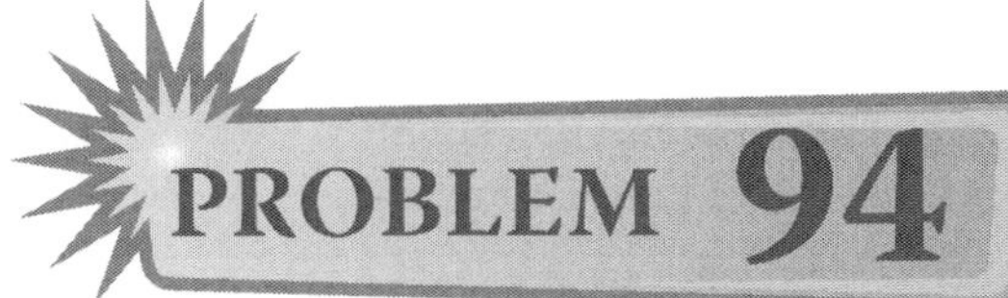

At a pet show, there were twice as many brown dogs as white dogs and 3 times as many grey dogs as white dogs. There was a total of 114 dogs. How many grey dogs were there?

Answer: ....

Mr. Fred is buying chocolates for the 4 dozen students of his class. The chocolates come in a packaged of 10 in a box. What is the least number of boxes he can buy so that each student gets at least 1 chocolate?

Answer: ....

Derrick had 168 stamps. He had 3 times as many Canada stamps as France stamps. His aunt gave him another 54 Canada stamps. How many Canada stamps did Derrick have in the end?

Answer: ....

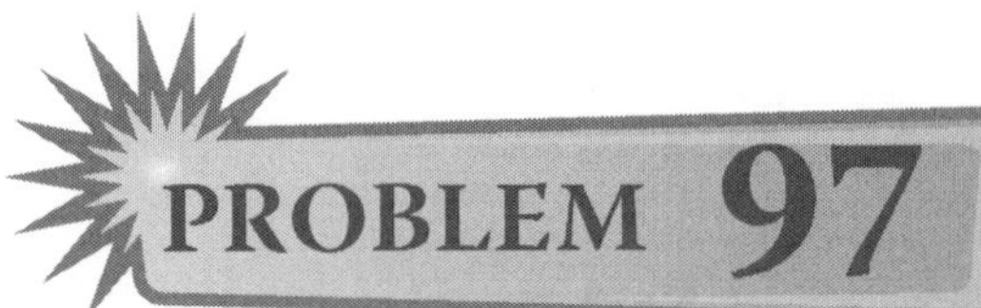

It takes 2000 bees one year to make 6 jars of honey. How many years will it take 4000 bees to make 60 jars of honey?

Answer: ....

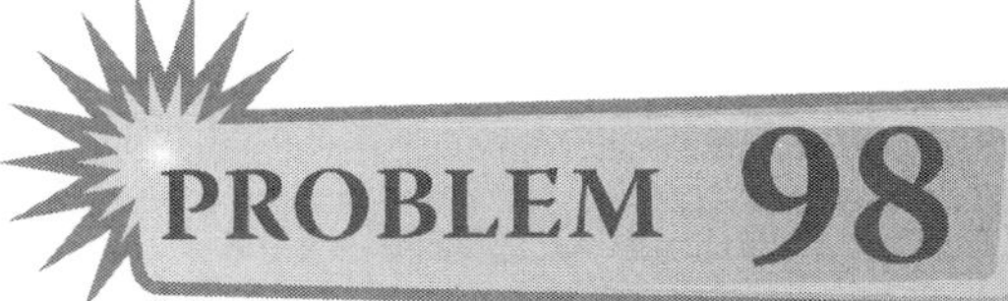

Some girls went for rock climbing. The troop leader asked 7 girls when they wanted to go rock climbing. Three of the 7 girls wanted to go rock climbing in the morning. If there were 28 girls on the trip, how many do you think wanted to go rock climbing in the morning?

Answer: ....

Liz went for fishing with her father at 7:00 A.M. There is a limit of catching 15 fish per hour at the fishing ground. If Liz needs 110 fish; by what time will she be able to catch her 110 fish?

Answer: ....

Linda has a dollar's worth of 20 cents and a dollar's worth of 10 cents. What is the ratio of 20 cents to 10 cents?

Answer: ....

PROBLEM 101

A novel with 320 pages has about 600 words per page. Liz read the book at a rate of 300 words a minute. In how many hours would she be able to finish the book?

Answer: ....

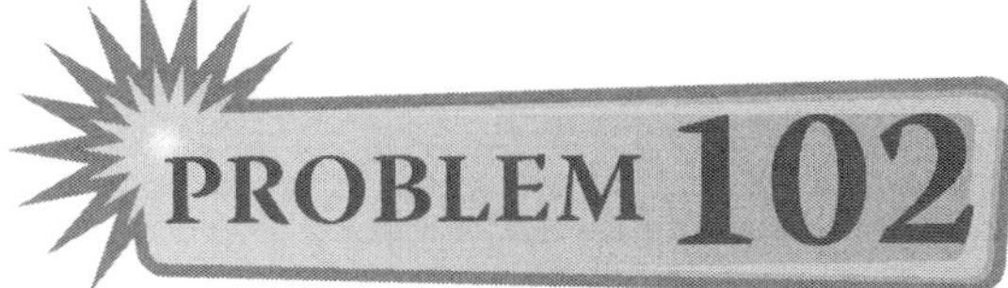

Laura sold twice as many chocolate cookies as cakes and three times as many lemon tarts as cakes. If Laura sold 27 lemon tarts, how many cookies did she sell?

Answer: ....

The sum of three numbers is 81 and their ratio is 3 : 7 : 17. What is the value of the smallest number?

Answer: ....

Martha practiced playing guitar on 12 different days in January. What is the ratio of the days that she practiced guitar to all the days in January? Write your answer in simplest form.

Answer: ....

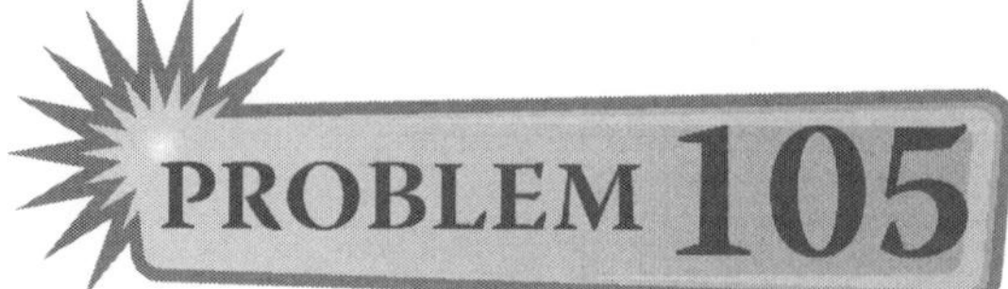

Jimmy can eat one sixth of a hot dog in two minutes. It takes 3 minutes for Lucas to eat one quarter of the hot dog. If Jimmy and Lucas start eating one hot dog each, who will finish first?

Answer: ....

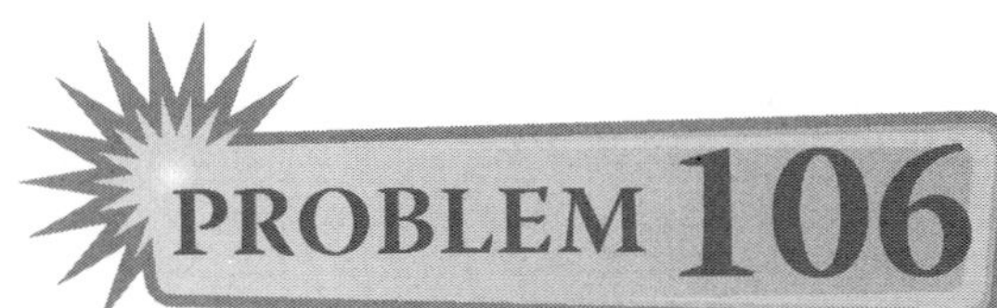

Kathy is 25 years old. Her sister is 4/5 of her age. The ratio of her sister's age to their mother's age is 4:10. How much older is Kathy's mother than her?

Answer: ....

Jenny performs ballet 2 days out of every 7 days. What is the ratio of days she performs to days she does not perform in a month? Write your answer in simplest form.
(Assume 1 month = 4 weeks = 28 days)

Answer: ....

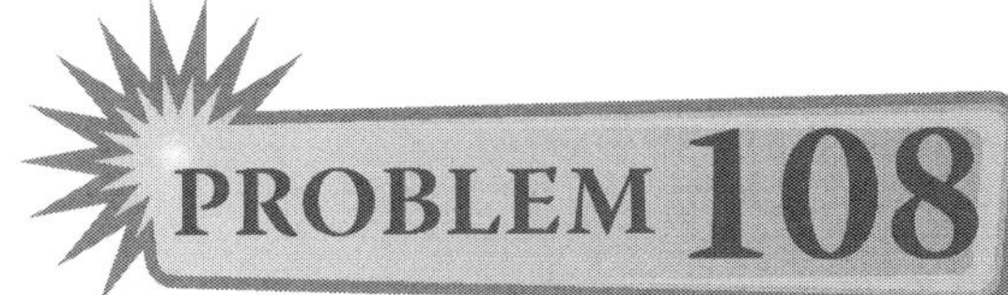

Physics tells us that weight of an object on the moon is proportional to its weight on Earth. Suppose an 80 kg man weighs 30 kg on the moon. What will a 60 kg man weigh on the moon?

Answer: ....

If Brian types 1 page each minute and Wendy types 3 pages each minute, in how many minutes will Brian be 100 pages behind Wendy?

Answer: ....

MONEY

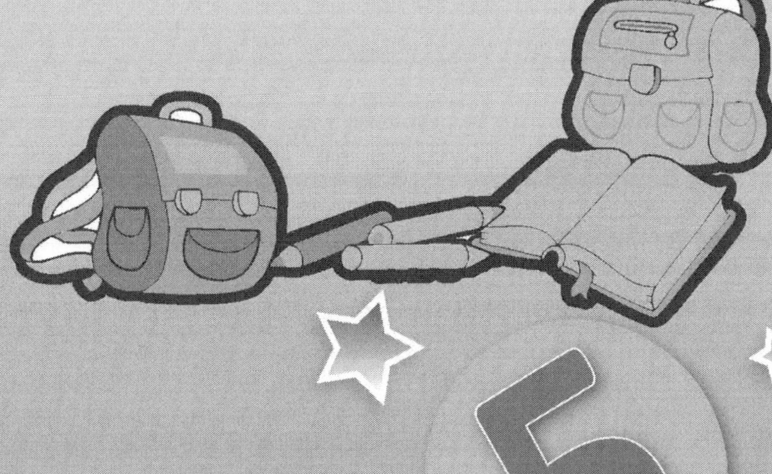

5

Guaranteed to improve children's
Math and success at school.

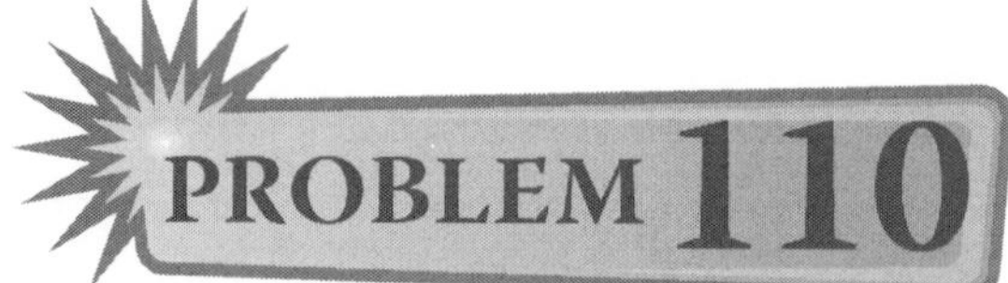

Mr. Ted bought a computer that was on sale for $910. The regular price of the computer was $1,495. How much amount of money did Mr. Ted save by buying the computer on sale?

Answer: ....

Pete paid for 4 identical pizzas with a $50 bill. If Pete got $3.60 in change, how much did one pizza cost?

Answer: ....

Sam had \$45. He spent all his money on a photo frame and a book. The book cost \$3 less than the photo frame.

a) How much did the book cost?

b) How much did the photo frame cost?

Answer: ....

Kenneth had $150 more than Jason and $70 less than Daniel. The three boys had a total of $790. How much did Daniel have?

Answer: ....

An apple and 2 bananas cost $2.70. The banana cost 30 cents less than the apple. How much did the apple cost?

Answer: ....

An adult concert ticket cost $15. A child concert ticket cost $7 less than the adult. Mr and Mrs James took some children to the concert. They paid a total of $86. How many children did they take to the concert?

Answer: ....

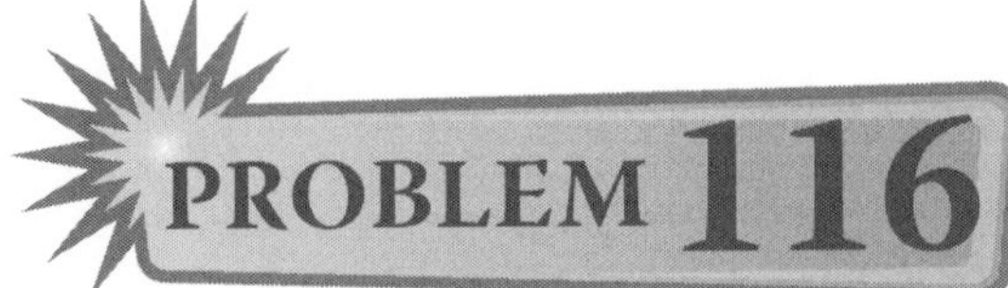

Janice had \$18.95. She bought a book that cost \$4.20, a toy that cost \$2.90 and a pair of socks. She had \$9.05 left. How much did she pay for the pair of socks?

Answer: ....

Bob's monthly salary in 2010 was $6800.
In 2011, he received a pay increase of 18%.
How much more did he earn in 2011 than in 2010?

Answer: ....

A sales tax of 7% is charged for all items sold at a shop. Jennifer paid $220 for a necklace. How much sales tax was charged for the necklace?

Answer: ....

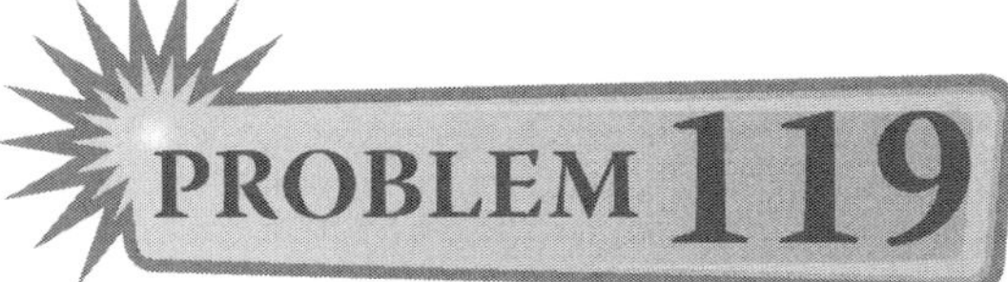

During a sales promotion a motorbike was on offer at 10% of its original price. After the promotion, the price increased by 5% of its original price. Jimmy bought the bike during the sale, paying $500 as down payment and the rest in 12 monthly installments of $350 each. How much was the motorbike after the sales promotion?

Answer: ....

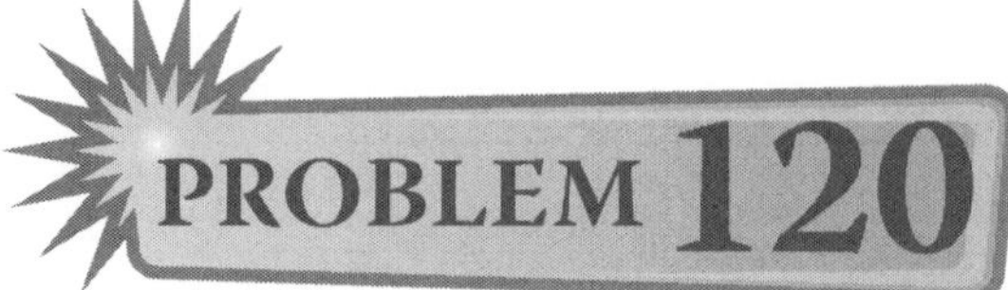

Lucy spent $60 on 120 pencils.

a) How much did she pay for each pencil?

b) A pen costs 25 cents more than a pencil. How many pens did she buy if she spent $60 to buy pens?

Answer: ....

Tim wants to purchase a new Television which costs $3043. He first needs to pay a down-payment of $1015. The remaining amount will be paid in monthly installments over 12 months. How much is each monthly installments?

Answer: ....

A pair of shoes costs half as much as a pair of trousers. If the total cost of 3 such similar shoes and a pair of trousers is $250, what is the cost of a pair shoes?

Answer: ....

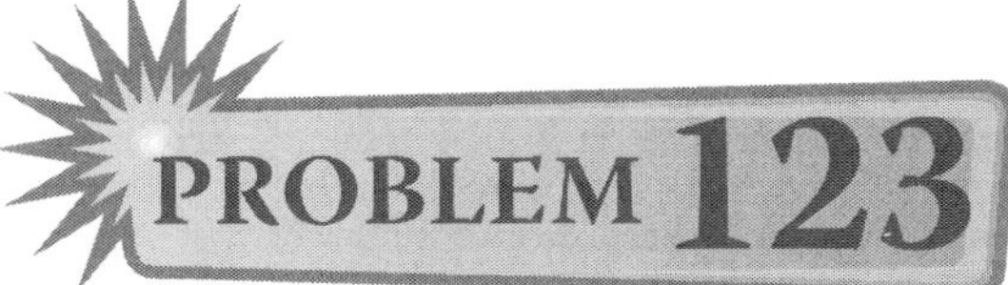

Samuel earns $2450 a month. He spends 2/5 of it and saves the rest. Annie earns $800 less than Samuel. Annie spends $180 more than Samuel each month. How much more can Samuel save than Annie in half a year?

Answer: ....

Sarah bought 9 buttons and ribbons for a total of $5.90. Each button costs $0.60, and each ribbon costs $0.70. How many buttons did Sarah buy?

Answer: ....

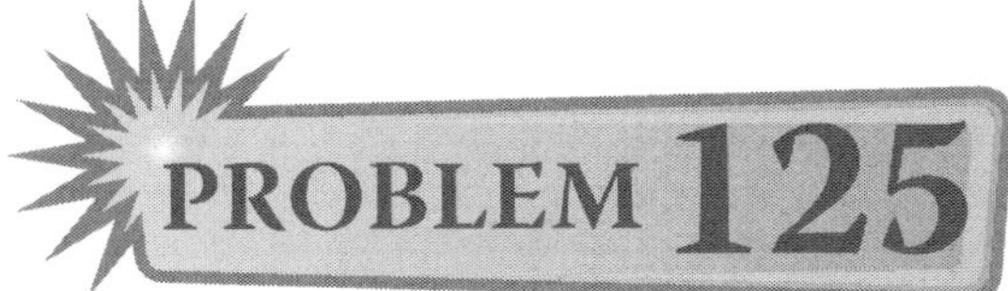

A potato and two oranges cost $2.30. Two potatoes and an orange cost $2.50. What is the cost of a potato and an orange?

Answer: ....

Peter has $4 in his piggy bank and John has $2 in his piggy bank. Peter adds $1 to his piggy bank each day and John adds $3 to his each day. On which day will John's saving be twice of Peter's?

Answer: ....

The cost of a novel and 2 books is $34.
The cost of 2 novels and 1 book is $86.
If Thomas buys 5 novels and 5 books, how much money does he spend?

Answer: ....

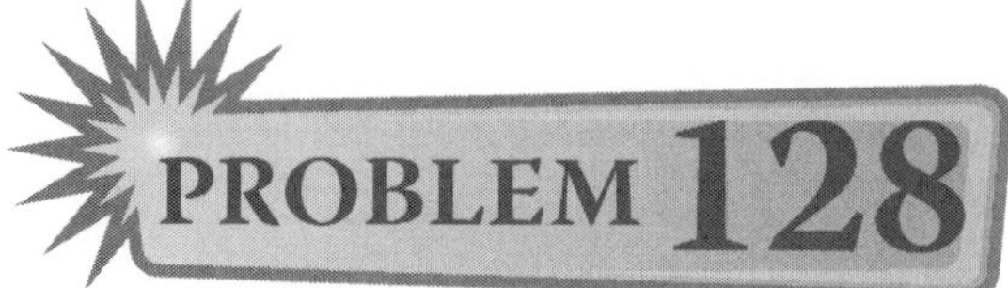

One plate costs thrice as much as a spoon. If 1 plate and 4 spoons cost $210. Find the cost of one spoon.

Answer: ....

The total cost of a coke can and an ice-cream cone is \$12. If the ice-cream cone cost \$3 more than the coke, how much does each item cost?

Answer: ....

A pencil and an eraser cost $1.1. If the pencil costs $1 more than the eraser, what is the cost of an eraser?

Answer: ....

George paid $15 for an equal number of note books and pens. Each notebook costs $0.80 and each pen costs one-fourth the cost of a notebook. How many pens did he buy?

Answer: ....

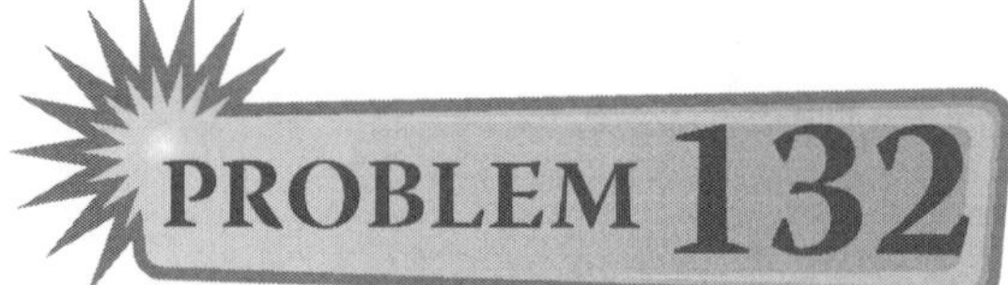

Jimmy had $55 in his wallet and his mother gave him some more money to go for shopping. He bought a pair of trousers at $31, two T-shirts at $18.00 each and 1 pair of shoes at $37. After the shopping, he had $32.00 left. How much money did Jimmy's mother give him?

Answer: ....

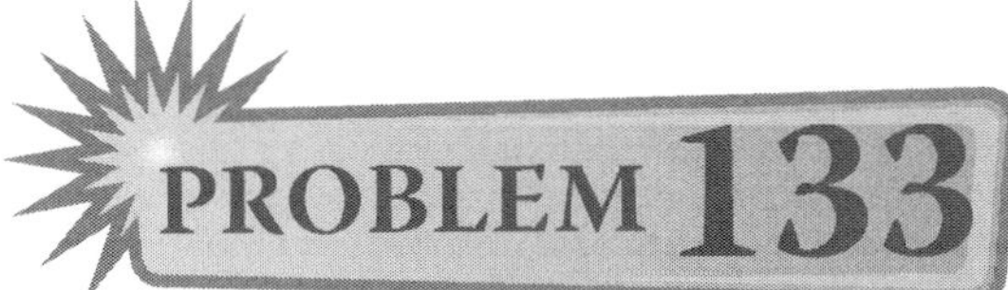

Ted has $500 in his savings. If his savings double in value every year for each of the next 5 years, how much money would he save at the end of 5 years?

Answer: ....

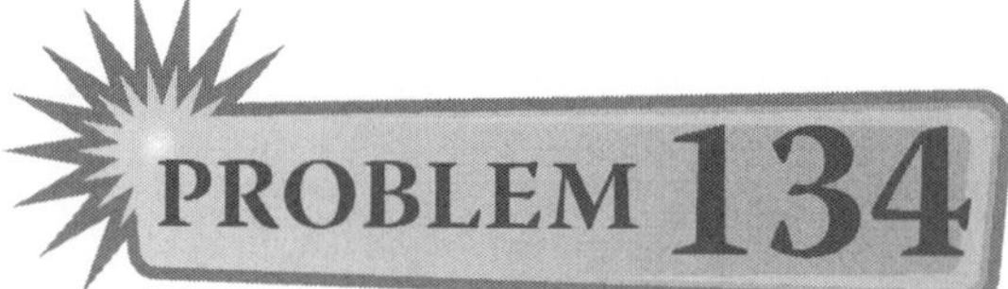

Mrs. Ali Marino has a gift shop. She buys 3 hand bags for $45 and sells them for $32 per piece. Last month, she sold every hand bag she bought, and made a profit of $221. How many hand bag did she buy?

Answer: ....

Tim bought 10 apples and his friend Joe bought 7 mangoes. The price of each apple is $0.55 and the price of each mango is $1.10. What is the total price of the fruits Tim and Joe bought together?

Answer: ....

Sarah needs to buy pencils for school. She can buy 12 pencils for $6 or 20 pencils for $7.50. Which is a better deal for Sarah to buy pencils?

Answer: ....

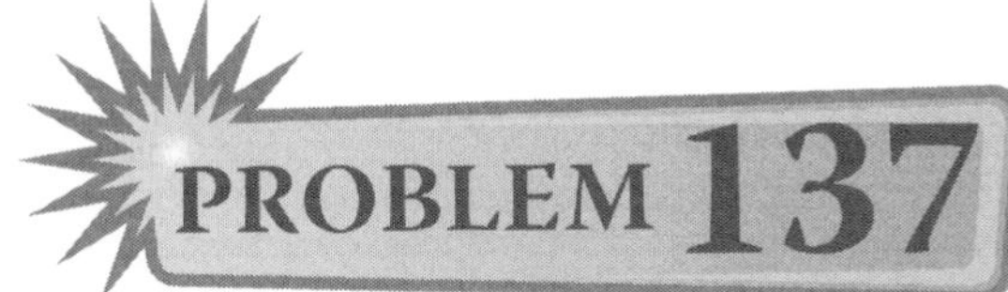

Jim bought a new bicycle at a 25% discount on its retail cost. The original price of the bicycle was $85. If Jim paid with a $100 bill, how much money did he receive back?

Answer: ....

Juliet bought 10 markers for $18. The markers she bought are red, blue and golden. Red markers are $1 each, blue markers are $2 each and golden markers are $5 each. If she bought at least one of each, how many red markers did she buy?

Answer: ....

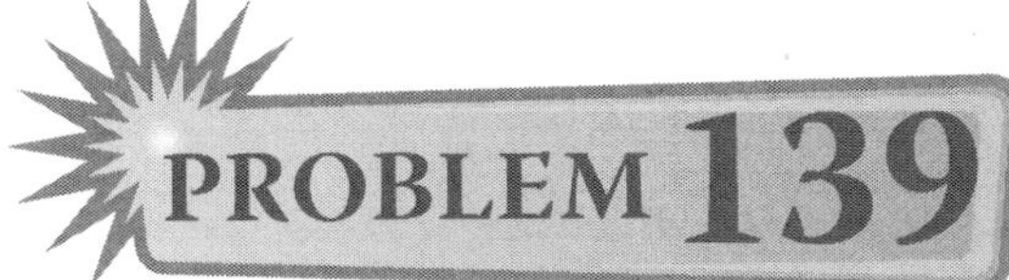

A painter charges $ 335 for materials and $ 25 per hour for his labour. The total cost of painting a house is $ 415. How many hours did it take the painter to paint the house?

Answer: ....

MEASUREMENT AND TIME

5

Guaranteed to improve children's math and success at school.

Tia made a sandwich with some cheese and 2 slices of bread that each weighed 2 grams. She weighed the sandwich and found it weighed 9 grams. How much cheese did she use?

Answer: ....

PROBLEM 141

Michael drinks a 350 ml bottle of lemon water every day. How much lemon water will he drink in one week? How much is this in liters?

Answer: ....

A car uses 3.5 liters of fuel every 2 kilometres it travels. How much fuel does it use if it travels 75 kilometres?

Answer: ....

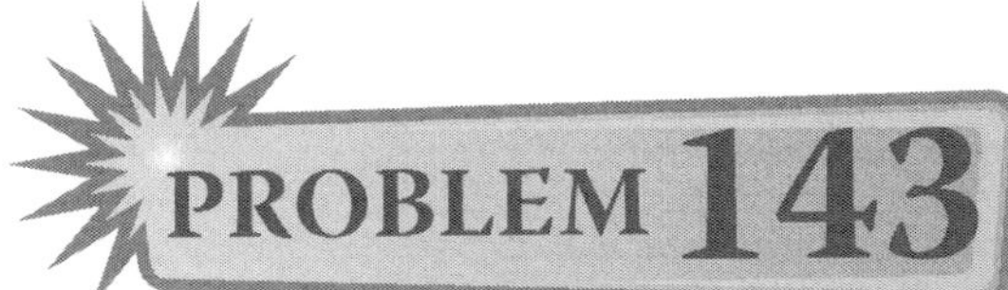

Amie has a jug of lemonade. She does not know how much lemonade she has, but she knows she can fill 12 glasses which have a capacity of 280 ml each. How much lemonade does she have?

Answer: ....

Jug A holds 1800 ml water. Jug B holds 3/4 more. How much water does jug B hold? How much water do the two jugs hold?

Answer: ....

PROBLEM 145

Sarah creates a Fruit punch. It contains 1/10 of a liter of apple juice, 2/5 of a liter of orange juice and 1/8 of a liter of grape juice. Which jug is the most suitable for Sarah to serve her fruit punch in?

Jug 1 0.4 liters	Jug 2 5 liters	Jug 3 755 ml

Answer: ....

Two students set their digital watches to 8:00 A.M. One watch runs one minute slow per hour, and the other watch runs 2 minutes fast per hour. What time will the slow watch show when it is exactly one hour behind the fast watch?

Answer: ....

PROBLEM 147

Mickey is twice as old as Donald and Donald is twice as old as Jerry. If Jerry is 12 years old, how old is Mickey?

Answer: ....

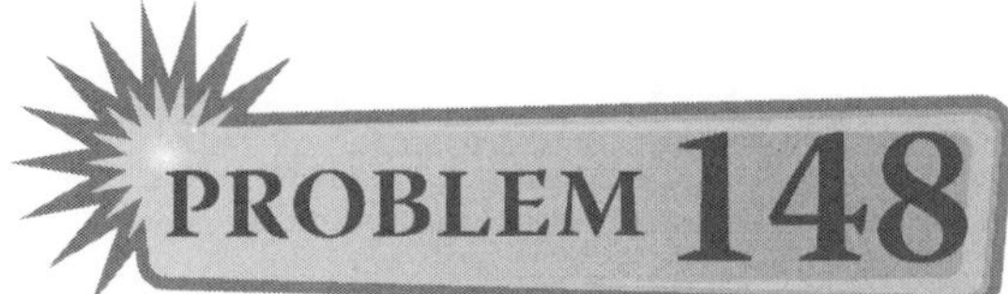

What is the correct time 3600 seconds before 1:30 P.M.?

Answer: ....

A planet far away in the space has 15 months in a year, 7 weeks in a month and 24 days in a week. A person on that planet is 7 years old. How old is he on earth?

Answer: ....

Nicolas is 4 years old and Susan is 24 years old. When will Susan be thrice as old as Nicolas?

Answer: ....

From 6 days 7 hours 35 minutes, subtract 3 days 9 hours 50 minutes.

Answer: ....

Mother's Day always falls on the first Monday of May. Stacy and Richard got married on Wednesday 18th May 2011. What was the date of Mother's Day in 2012?

Answer: ....

SPEED, DISTANCE AND TIME

5

Guaranteed to improve children's Math and success at school.

Jimmy dropped a ball from a height of 250 meters to the ground. The ball rebounded back to the height from which it had been dropped each time it hit the ground. Find the total distance the ball had traveled in the air when it hit the ground for the 4th time.

Answer: ....

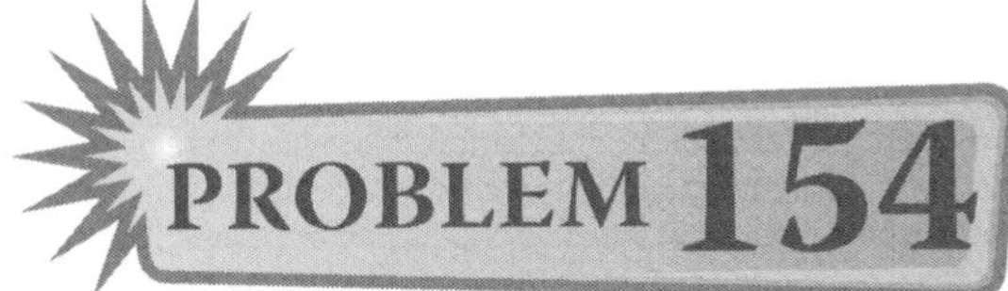

Machine A can make 2050 ice-creams a day. This is 110 fewer ice-creams than what machine B can make. 10 ice-creams are packed in a packet and every 1000 packets of ice-creams are placed in a box.

a) If both machines A and B are used, how many packets of ice-creams will be there after a week?

b) How many complete boxes of ice-creams will there be after a week?

Answer: ....

Jimmy is making snowballs for his snow castle. He makes 18 snowballs in one hour, but 4 snowballs melt every 20 minutes. How long will it take to make 250 snowballs?

Answer: ....

PROBLEM 156

Jane eats twice as many sweets as Shelly in half the time. Shelly eats 12 sweets in 4 minutes. How many sweets does Jane eat in the same time?

Answer: ....

Tim ran 200 miles in 10 hours, walked 40 miles in 20 hours, and then ran another 360 miles in 10 hours. What was his average speed for the whole trip in miles per hour?

Answer: ....

PROBLEM 158

A bus traveling at an average rate of 50 kilometers per hour made the trip to town in 6 hours. If it traveled at 45 kilometers per hour, how many more minutes would it take to make the trip?

Answer: ....

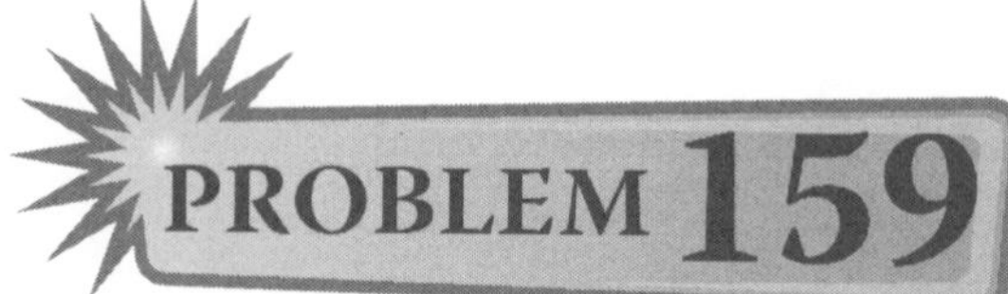

A bus and a car leave the same place and travel in opposite directions. If the bus is traveling at 50 kilometers per hour and the car is traveling at 55 kilometers per hour. In how many hours will they be 200 kilometers apart?

Car ←——o——→ Bus
200

Answer: ....

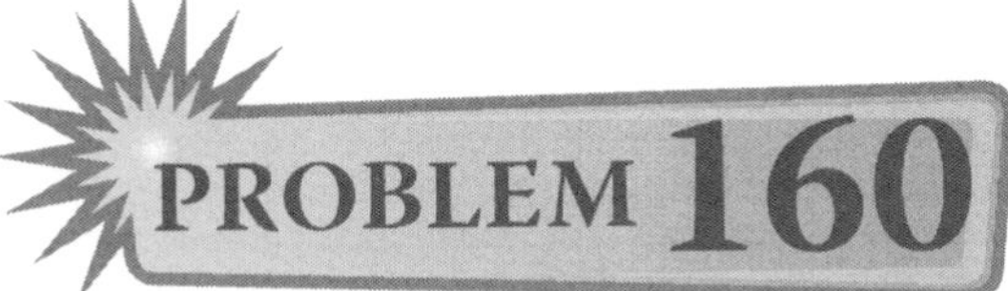

John took a drive to the town at an average rate of 40 kilometers per hour. In the evening, he drove back at 30 kilometers per hour. If he spent a total of 6 hours traveling, what is the distance traveled by John?

Answer: ....

Lily and Trevor start running in opposite directions. Lily runs 1 meter per second and Trevor runs 2 meters per second. How far apart are they in kilometers after 2 hours?

Lily ←——o——→ Trevor

Answer: ....

PROBLEM 162

At 11:00 A.M., a yellow car leaves city "A" at a constant rate of 60 km/hr. towards city "B". At the same time a blue car leaves city "B" toward city "A" at the constant rate of 50 km/hr. The distance between cities A and B is 240 kilometres. At what time will the two cars cross each other?

A ← X ○ 240 – X → B

Yellow car   Blue car

Answer: ....

A total distance of 650 kilometres was covered by a plane in 6 hours at two different speeds. For the first part of the trip, the average speed was 105 kilometres per hour. The remainder of the trip was flown at an average speed of 110 kilometres per hour. For how long did the plane fly at each speed?

Answer: ....

PROBLEM 164

Raymond drove from home at an average speed of 30 kilometers per hour to the airport. He boarded an airplane that flew to Paris at an average speed of 60 kilometers per hour. The entire distance was 300 kilometers and the entire trip took six hours. Find the distance from the airport to Paris.

Answer: ....

An airplane made a trip to Las Vegas and back. While going to Las Vegas the airplane flew at a speed of 430 kilometers per hour and on the return trip it flew at a speed of 460 kilometers per hour. How long did the trip take if the return trip took 6 hours?

Answer: ....

Given the piece of paper shown below that can be folded up to form a cube, what numbered face will be opposite the number 6 face when folded?

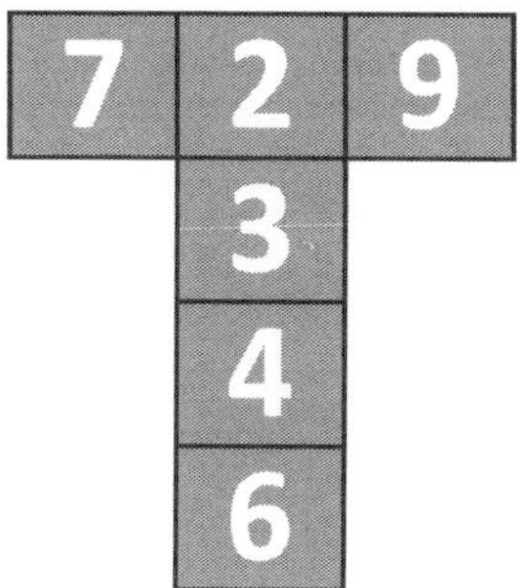

Answer: ....

AREA, PERIMETER AND VOLUME

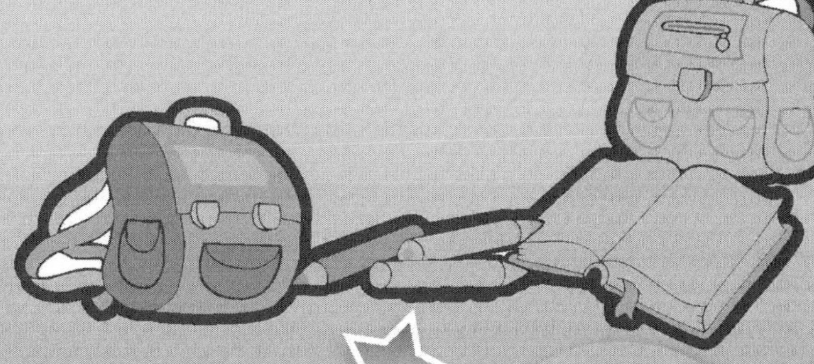

5

Guaranteed to improve children's math and success at school.

What is the area of the total shaded portions of the rectangles shown below?

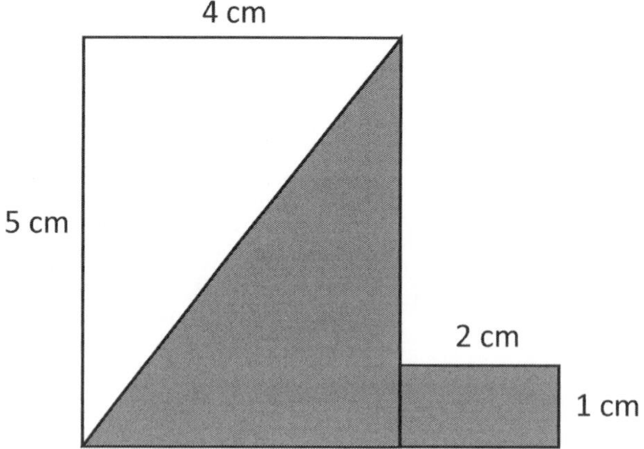

Answer: ....

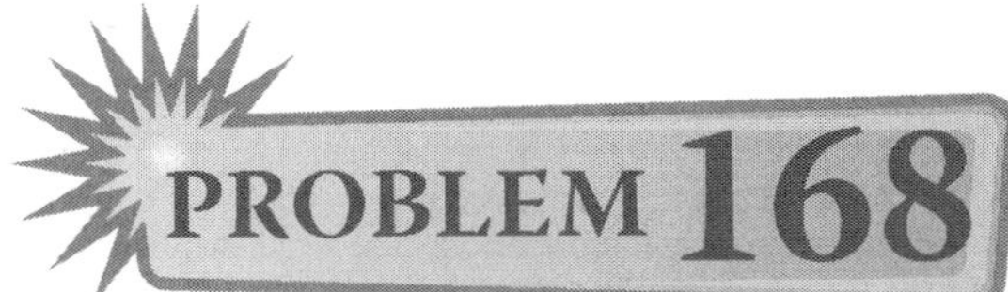

Raymond's bedroom door is 9 feet tall and 4 feet wide. What would be the cost of a new door if the carpenter charges $15 per square foot to make it?

Answer: ....

Find the length of side AB and then find the area of the shape?

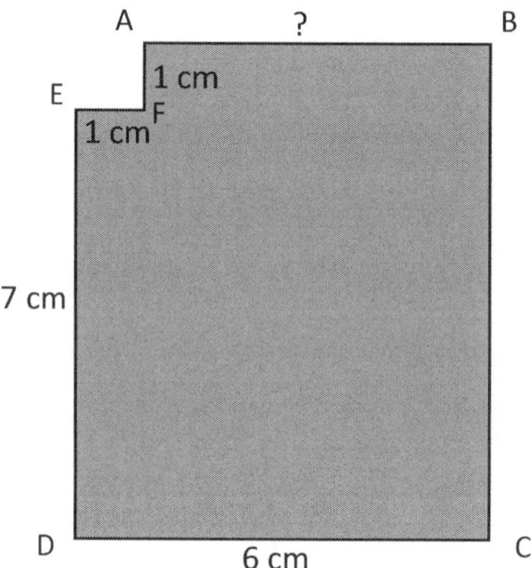

Answer: ....

A picture of a house is shown below. The shaded portion at the bottom represents the carpet. What is the area of the carpet?

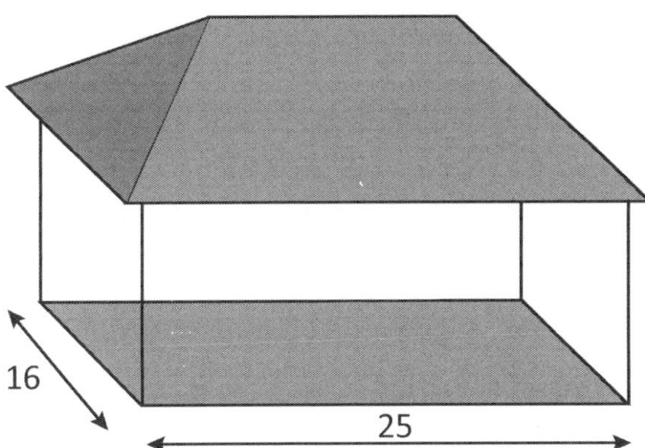

Answer: ....

What is the area of the shape?

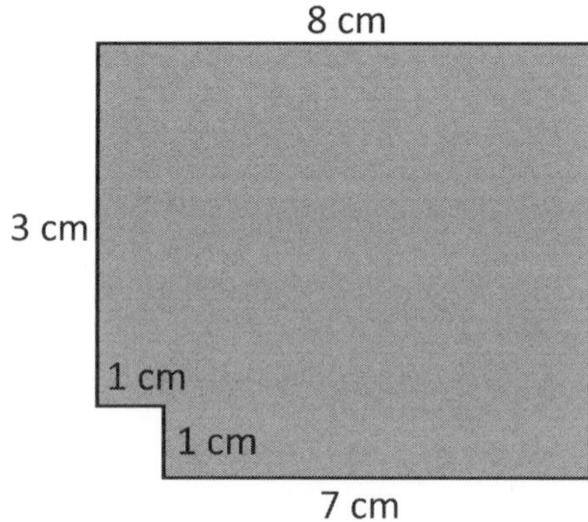

Answer: ....

PROBLEM 172

The figure below shows two squares with sides 6cm and 12 cm.

a) Find the ratio of the perimeter of the smaller square to the perimeter of the bigger square.

b) What is the area of the shaded part of the figure?

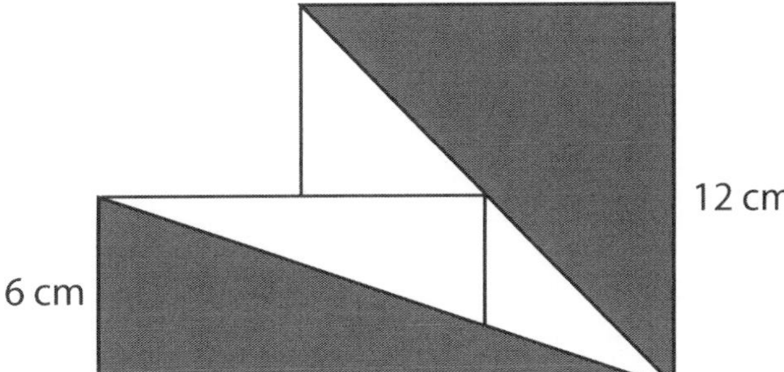

Answer: ....

Find the area of the shaded part of the figure shown below.

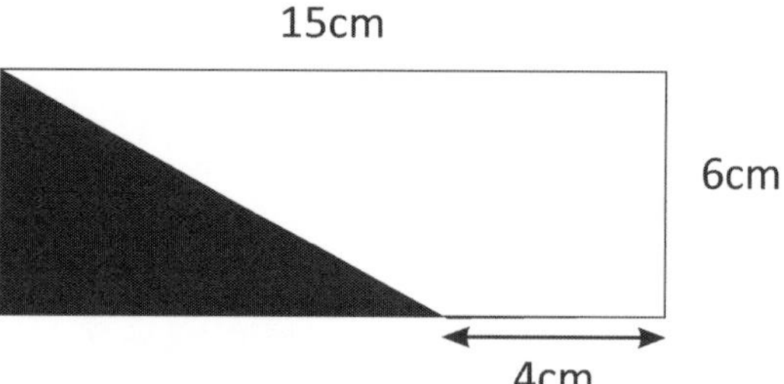

Answer: ....

PROBLEM 174

The area of rectangle ABCD is 117 meter square. Find the length and perimeter.

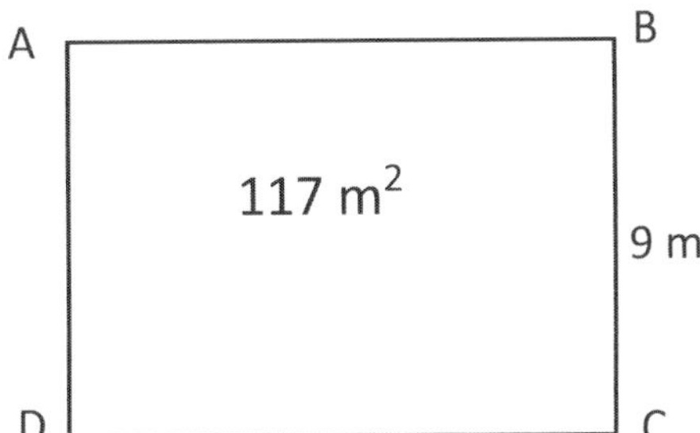

Answer: ....

A wooden block measures 30 cm by 26 cm by 10 cm. Find the greatest number of 2-cm cubes that can be cut from it.

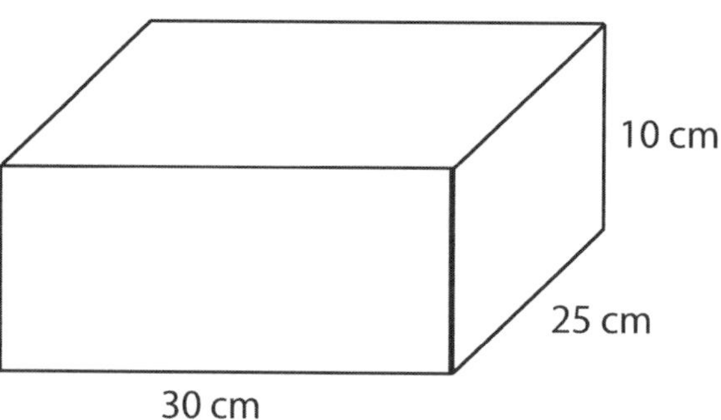

Answer: ....

PROBLEM 176

The sides of a triangle are in the ratio 4:5:6. If the shortest side is 6 centimeters, what is the perimeter of the triangle in centimeters?

Answer: ....

If the perimeter of a rectangle is 20 centimeters and the ratio of it length to its width is 1: 4, what is the area of the rectangle?

Answer: ....

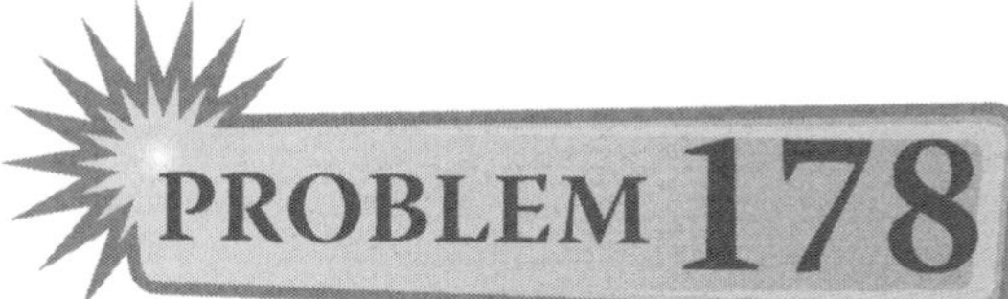

A farmer has a piece of land that measures 65m by 50m. In order to fence it, he puts stakes all around the land that are 5 meters apart. How many stakes will he need in all to fence his land?

Answer: ....

PROBLEM 179

ABCD is a square. Find the total area of the shaded portion.

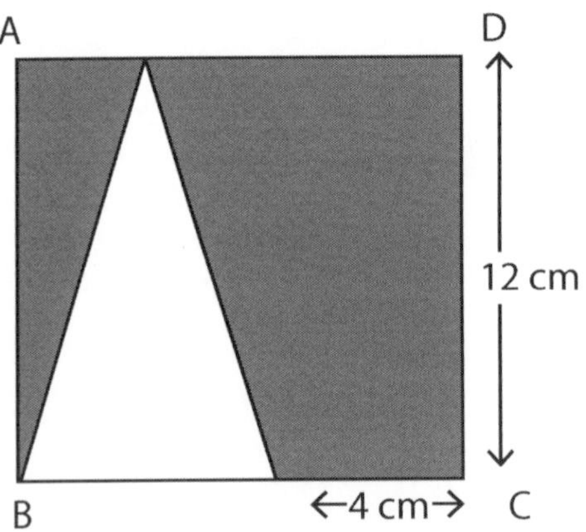

Answer: ....

What value of x is needed to make the figures similar?

5 cm

7 cm

10 cm

X

Answer: ....

The side of the square is 8 centimeters and the base of the triangle is 5 centimeters. What is the area of the unshaded region in centimeters?

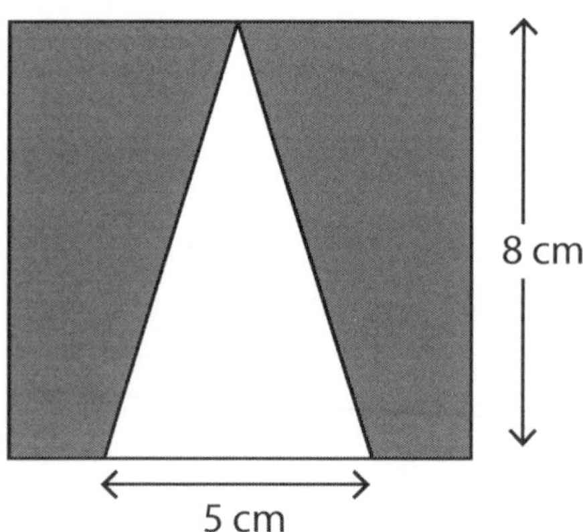

Answer: ....

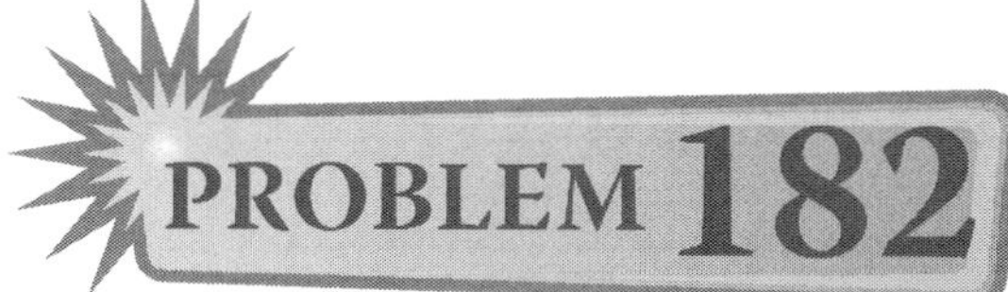

My school has a swimming pool 30 meters long and 10 meters wide. It is surrounded by a grass lawn that is 50 meters long and 20 meters wide as shown in the picture below. Find the area occupied by the grass?

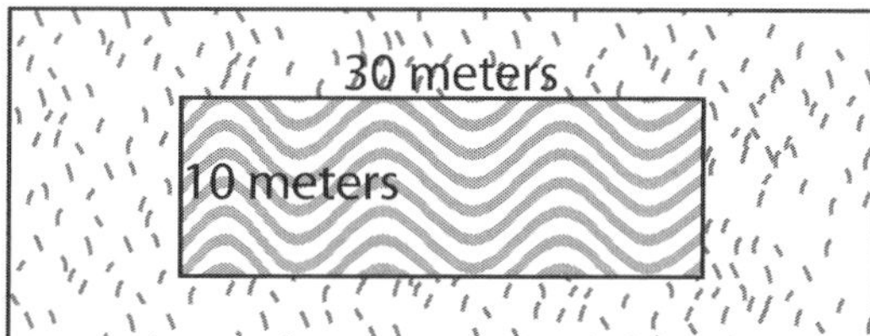

Answer: ....

The radius of circle A has the same length as the diameter of circle B. What is the ratio of circle A's area to circle B's area?

Answer: ....

PROBLEM 184

What is the fewest number of squares, each with a perimeter of 8 centimeters that would completely cover a square with a side of 8 centimeters?

Answer: ....

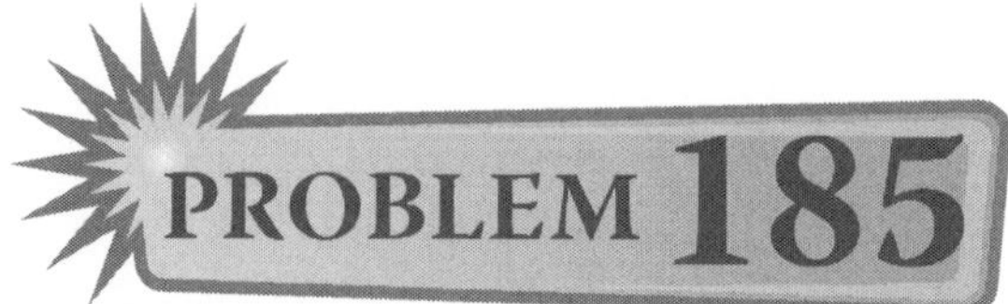

The length of a rectangle is 130% that of its width. The perimeter of the rectangle is 92cm. Find the area of the rectangle.

Answer: ....

Nicole cuts four congruent squares from the corners of a wooden board that measures 15 centimeters long and 10 centimeters wide. If the equal squares measure 3cm on each side ; what is the area of the remaining cardboard.

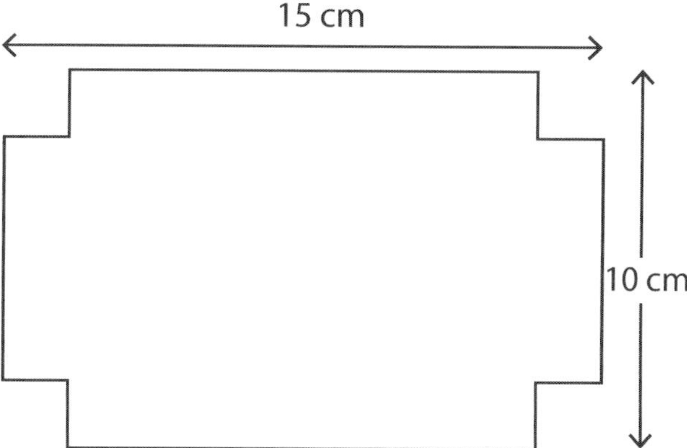

Answer: ....

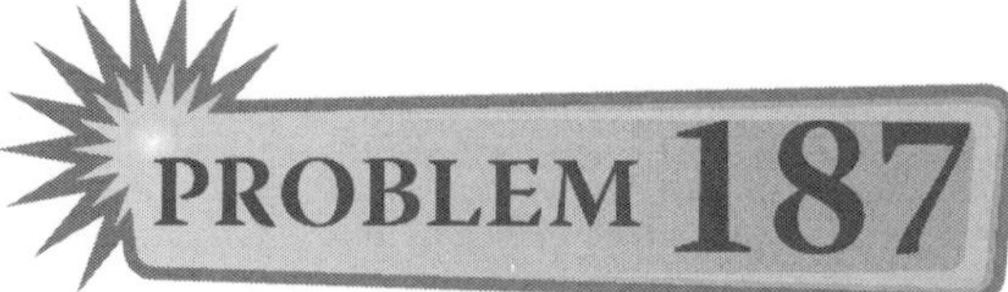

The perimeter of square A is 4 times the perimeter of square B. What is the ratio of the area of square A to the area of square B?

Answer: ....

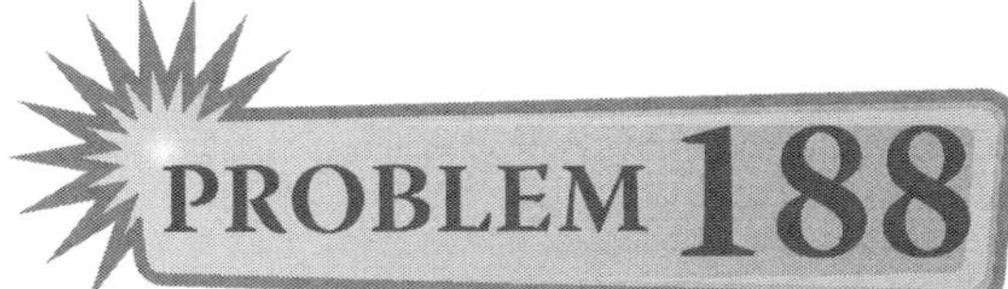

Sandra enlarged a picture proportionally. Her original picture is 4 cm wide and 6 cm long. If the new larger picture is 10 cm wide, what is its length?

Answer: ....

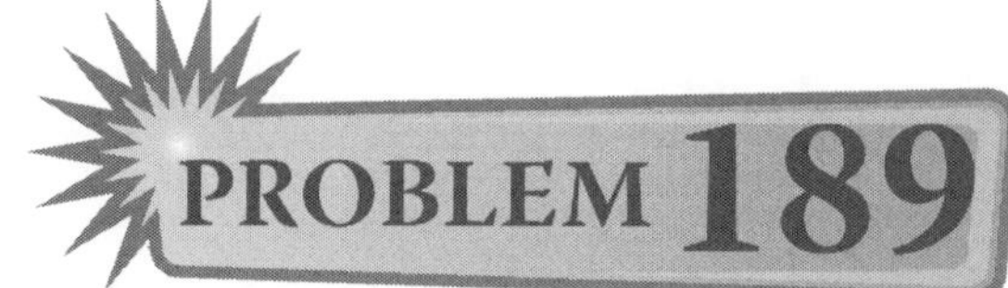

If the perimeter of square B is 2 times the perimeter of square A, what is the ratio of the area of square A to the area of square B?

Answer: ....

PROBLEM 190

The perimeter of a square is half the area of a rectangle.

A wire is bent to form 3 such squares and 2 such rectangles. There is a remainder of 20 cm of wire left.

If the length of the rectangle is 15 cm and its width is 12 cm, find the length of the wire.

Answer: ....

A wire is divided into 2 parts equally. One of the parts is bent to form two similar squares with sides 15 cm. The other part is divided into three similar rectangles with width 5 cm,

a) Find the perimeter of each rectangle

b) Find the area of each rectangle.

Answer: ....

PROBLEM 192

The figure shows a rectangle of width 9 cm and a square inside the rectangle of side 4 cm. The area of the shaded part is 83 cm^2.

a) Find the area of the whole rectangle.

b) Find the perimeter of the rectangle.

Answer: ....

The square below has an area of 64 m^2. The area of the rectangle is thrice as

much as that of the square shown below. Find the length of the rectangle.

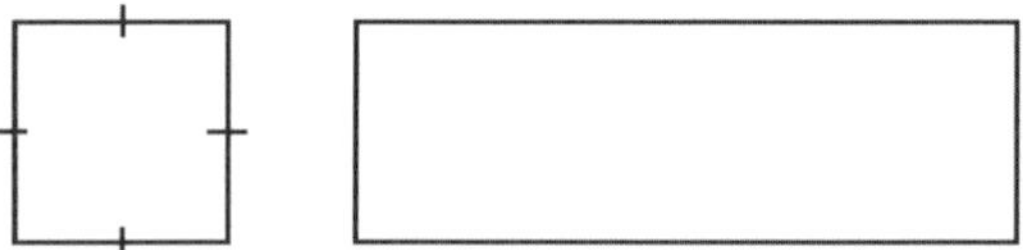

Answer: ....

The length of a field is twice its width. The perimeter of the field is 180 m. What is the total length of the field?

Answer: ....

The figure below is made up of 4 identical squares and was made with a wire of length 126 cms long. What is the perimeter of this figure?

Answer: ....

Each of the figures below is made up of identical small cubes , each of side 2 units.

What is the volume of figure 1 and figure 3, subtracted from the volume of figure 2?

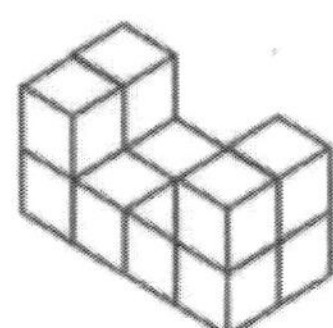

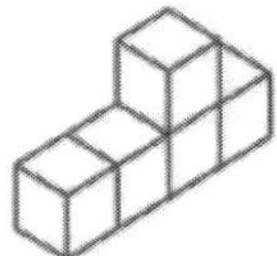

Answer: ....

3/4 of a rectangular tank with a base area of $800cm^2$ is filled with water. Another 2.8 liters of water is needed to fill this tank to its brim. What is the height of the tank?

Answer: ....

PROBLEM 198

A container measuring 7 cm by 10 cm by 22 cm is completely filled with juice. After some juice was poured into the glasses, the depth of the juice decreased to 15 cm. How much juice was poured out?

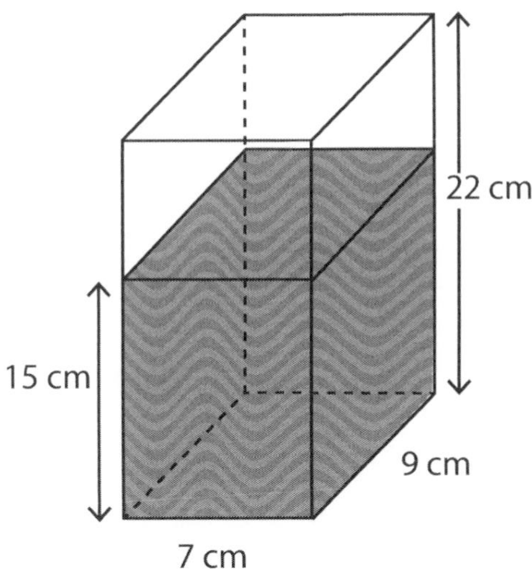

Answer: ....

A rectangular tank 50 cm long, 40 cm wide and 25 cm high was 3/4full of water. Water began to leak from the tank at 400 cm^3 per minute. At the same time a tap from which water is flowing at a rate of 2.4 liters per minute was opened into the leaking container. How long did it take to fill the tank completely with water?

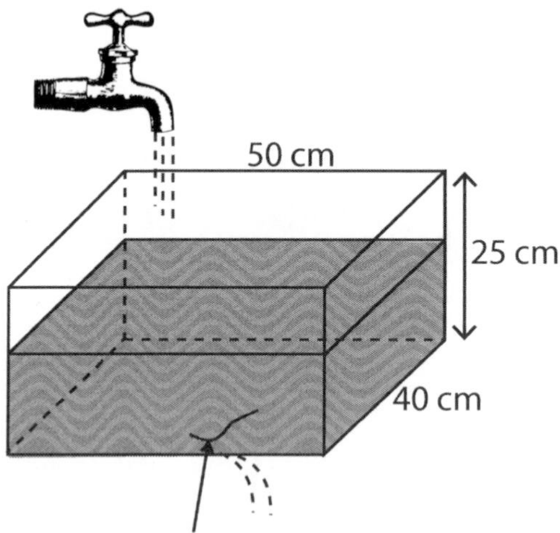

Answer: ....

The volume of water in the rectangular tank is 72 cm3.How much more water is needed to fill up the rectangular tank completely?

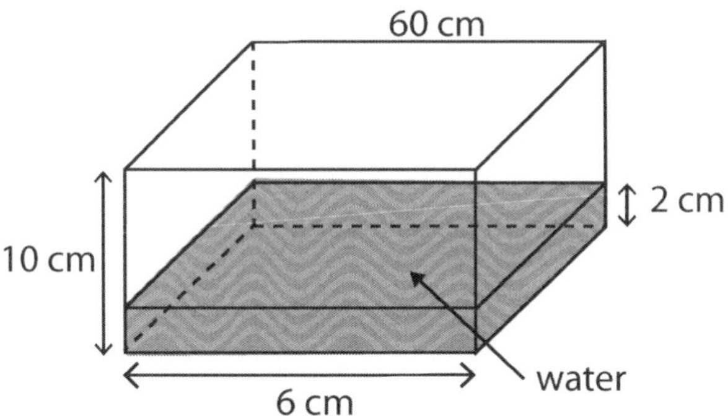

Answer: ....

DETAILED SOLUTIONS

5

Guaranteed to improve children's Math and success at school.

1. Numbers

Solution to Question 1

Prime number between 50 & 70 are
53, 59, 61, 67
Number of prime number less than 70 are 15 + 4 = 19

Solution to Question 2

Prime numbers below 100 are:

2	3	5	7	11	13	17	19	23
29	31	37	41	43	47	53	59	61
67	71	73	79	83	89	97		

Number of prime numbers between 1 to 100 are 25

Percentage of the numbers from 1 to 100 that are prime = 25/100 * 100 = 25%

Solution to Question 3

Following is the list of palindromes between 100 and 1000

101 - 111 - 121 - 131 - 141 - 151 - 161 - 171 - 181 - 191
202 - 212 - 222 - 231 - 242 - 252 - 262 - 272 - 282 - 292
303 - 313 - 323 - 333 - 343 - 353 - 363 - 373 - 383 - 393
404 - 414 - 424 - 434 - 444 - 454 - 464 - 474 - 484 - 494
505 - 515 - 525 - 535 - 545 - 555 - 565 - 575 - 585 - 595
606 - 616 - 626 - 636 - 646 - 656 - 666 - 676 - 686 - 696
707 - 717 - 727 - 737 - 747 - 757 - 767 - 777 - 787 - 797
808 - 818 - 828 - 838 - 848 - 858 - 868 - 878 - 888 - 898
909 - 919 - 929 - 939 - 949 - 959 - 969 - 979 - 989 - 999

There are 90 palindromes between 100 and 1000.
Instead of listing down all of them, smarter way is to list them down from 100 to 200 and then multiply them by 9 as there are 9 similar sequences, from 100 to 200, 201 to 300 and so on.

Solution to Question 4

Prime numbers below 100 are:

(2)	3	5	7	11	13	17
19	(23)	(29)	31	37	41	43
47	53	59	61	67	71	73
79	83	89	97			

Number of times the digit 2 appears in the prime numbers below 100 is 3

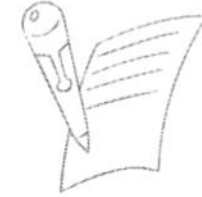

Solution to Question 5

Greatest 3 digit no = 999
Greatest 4 digit no = 9999
The sum of the greatest 3-digit number and the greatest 4-digit number
= 999 + 9999 = 10998
The difference between the greatest 4-digit number and the greatest 3-digit number = 9999 – 999 = 9000

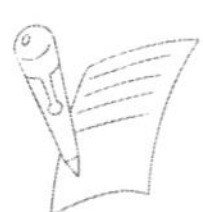

Solution to Question 6

The first perfect square 1, is the square of 1.
The second perfect square 4, is the square of 2.
The third perfect square 9, is the square of 3.
And so on....
So the 100th perfect square will be the square of 100
So the 100th perfect square = 100 x 100 = 10000

Solution to Question 7

I am a number between 190 and 207.
Multiples of 4 between 190 & 207 are:
192, 196, 200, 204
When divided by 4, the remainder is 2
Therefore the numbers can be above multiples + 2 =
194, 198, 202, 206
Multiples of 5 between 190 & 207 are:
195, 200, 205
When divided by 5, the remainder is 1.
Therefore the numbers can be the above multiples + 1
196, 201, 206
The common number in both the lists is 206
The number is 206.

Solution to Question 8

Multiples of 5 below 100 are
= 5 ,10, 15, 20, 25, 30, 35, 40, 45, 50, 55, 60, 65, 70, 75, 80, 85, 90, 95

Multiples of 7 below 100 are
= 7, 14, 21, 28, 35, 42, 49, 56, 63, 70, 77, 84, 91, 98

But both the lists will have common numbers that are divisible by both 5 and 7.
The LCM of 5 and 7 is 35.

So the multiples of 35 between 1 and 100 will be repeated in both the lists.
Multiples of 35 below 100 are 35 and 70.

So the number of positive integers less than 100 that can be divided by 5 or 7 = 31
(because 2 numbers are repeating in the lists above)

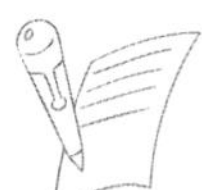

Solution to Question 9

No is divisible by 4 when the last two individual digits of the number are evenly divisible by 4.
So only the numbers ending in 48 and 84 will be divisible by 4.
All the four-digit numbers using each of the digits 4, 5, 8, and 9 exactly once are
5948, 5984, 9548, 9584,
Number that are multiple of 4 = 4

Solution to Question 10

Multiples of 4 between 190 & 207 are:
192, 196, 200, 204
When divided by 4, the remainder is 2.
Therefore the possible numbers are
194, 198, 202, 206
Multiples of 5 between 190 & 207 are:
195, 200, 205
When divided by 5, the remainder is 1.
Therefore the possible numbers are
196, 201, 206
The common number for both the conditions to be true = 206
The number is 206.

Solution to Question 11

The digits of the number add up to 12.
And the hundreds digit is two more than the ones (units digit).
The rule of divisibility by 3 is that the sum of digits should be divisible by 3. Since the total is 12, it will always be true.
Let us generate all possible such numbers by taking different values of the ones digit, putting the hundreds digit as two more than that and subtracting the difference from 12 of both these digits to get the tens digit.

If ones digit = 1, hundreds digit = 3, ten's digit = 12 – 1 – 3 = 8. Number = 381
If ones digit = 2, hundreds digit = 4, ten's digit = 12 – 2 – 4 = 6. Number = 462
If ones digit = 3, hundreds digit = 5, ten's digit = 12 – 3 – 5 = 4. Number = 543
If ones digit = 4, hundreds digit = 6, ten's digit = 12 – 4 – 6 = 2. Number = 624
If ones digit = 5, hundreds digit = 7, ten's digit = 12 – 5 – 7 = 0. Number = 705

If we take ones digit as more than 5, say 6, then the hundreds digit becomes 8 and the tens digit will become negative (12 – 6 – 8 = –2) which is not possible.
So there are 5 possibilities.

One more condition is that the digits have to be in descending order. Of these 5 possibilities, the only number that satisfies that condition is 543.

The secret number is 543.

Solution to Question 12

Divisibility rule for 9 is that the sum of all the digits in the number should be divisible by 9
So, 2 + 4 + 3 + 5 + X + 9 should be divisible by 9
Which means 23 + X should be divisible by 9

The multiple of 9 that is closest to and bigger than 23 is 27.
So the value of X is 4 and the number is
243549

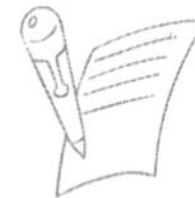

Solution to Question 13

Square of 1 is 1 and the square of 2 is 4. So the smallest number that has a square between 2 and 999 is 2.
At the other end,
302 = 900
312 = 961
322 = 1024
So the largest number that has a square and is less than 999 is 31.
The number of the integers from 2 to 999 that are squares of a whole number =
31 – 1 = 30

Solution to Question 14

The average of 7 consecutive numbers is 2009.
If the first number is X, then the other numbers will be X + 1, X + 2, X + 3, X + 4, X + 5 and X + 6
The average = 2009
(X + X + 1 + X + 2 + X + 3 + X + 4 + X + 5 + X + 6) / 7 = 2009
7X + 21 = 2009 x 7
7X = 14063 – 21
X = 14042/7
X = 2006
The first number is 2006
The average of the first three of these numbers = (2006 + 2007 + 2008)/3
= 6021/3
= 2007

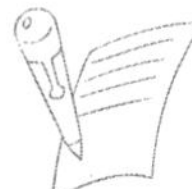

Solution to Question 15

Dividing 11		Dividing 21	
Divisor	Remainder	Divisor	Remainder
2	1	2	1
3	2	3	0
4	3	4	1
5	1	5	1
6	5	6	3
7	4	7	0
8	3	8	5
9	2	9	3
10	1	10	1

The largest number that gives the same remainder when dividing 11 and 21 is 10.

Solution to Question 16

Number = Divisor x quotient + remainder
= 205 * 76 + 13
= 15580 + 13
= 15593
The number is 15593.

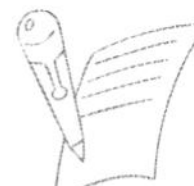

Solution to Question 17

77235 ÷ 7=	Quotient = 11033	Remainder = 4
11111 ÷3	Quotient = 3703	Remainder = 2
10001 ÷2	Quotient = 5000	Remainder = 1

Sum of remainders = 4 + 2 + 1 = 7
When 7 is divided by 7, remainder = 0

Solution to Question 18

If X is the first number then 120 – X is the other number
(1/6)X =(1/2)(120 – X) +8
X/6 = 60 – X/2 + 8
X/6 + X/2 = 68
4X/6 = 68
X = 68 x 6 / 4 = 102
The numbers are 102 and 18.

Solution to Question 19

There are several ways to do this and a few have been mentioned below:

1 2 + 3 – 4 + 5 + 6 7 + 8 + 9 = 100

1 + 2 + 34 – 5 + 67 – 8 + 9 = 100

123 – 4 – 5 – 6 – 7 + 8 – 9 = 100

12 x 3 – 4 – 5 – 6 + 7 + 8 x 9

12 x 3 – 4 x 5 + 67 + 8 + 9

12 ÷ 3 + 4 x 5 – 6 – 7 + 89

12 ÷ 3 + 4 x 5 x 6 – 7 – 8 – 9

Solution to Question 20

Let the first odd no be X. Since consecutive odd numbers differ by 2, the next numbers will be X + 2, X + 4 and X + 6
X + X + 2 + X + 4 +X + 6 = 88
4X + 12 = 88
4X = 88 – 12
X = 76/4 = 19
4 consecutive odd no's are 19, 21, 23 and 25.

Solution to Question 21

Let the two numbers be X and Y.
X + Y = 15
The difference between the same two numbers is 3.
X – Y = 3
X + Y = 15
Adding both the equations we get
2X = 18
X = 18/2 = 9
Y = 9 – 3 = 6
The two numbers are 9 and 6.

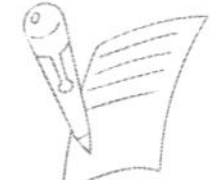

Solution to Question 22

One of the numbers is half of the other.
If one number is X, the other number will be 2X.
When two numbers are added, the sum is 27.
2X + X = 27
3X = 27
X = 27/3 = 9
The two numbers are 9 and 18.

Solution to Question 23

As the boat can carry 150 kg then there is only one choice i.e. the 60 kg friend has to go and drop the 80 kg friend because their total weight will be
80 + 60 = 140 kg which satisfies the given condition.
Two people weighing 80 kg cannot go at the same time.
1: 60kg + 80 kg — go across
2: 60 Kg comes back
3: 60kg + 80 kg — go across
4: 60 Kg comes back
5: 60 kg + 80 kg go across — go across
So 5 trips across the river are needed to get the 4 friends over to the other side.

Average time = (65 + 57 + 56 + 59 + 66) / 5
= 303/5 = 60.5 minutes
The mean (average) number of minutes it took Tracy to drive from her house to her office = 60.5 minutes

Solution to Question 25

Sarah's average score in 5 tests is 98.
That means her total score in the 5 tests is 98 x 5 = 490.
The maximum score she can get in 5 tests is 500 (if she gets 100 in all 5).
Since she got either 90 or 100 in each test, that means she got only one 90 and rest all 100.
If she had got more than one 90, her score would have been less than 490.
Therefore she got one 90 and four 100s in her 5 tests

Solution to Question 26

17 of the balls are not green , so 3 balls are green
5 are black
12 are not yellow , so 8 are yellow
Therefore number of blue balls Kevin has = 20 – 3 – 8 – 5 = 4

Solution to Question 27

No of palm trees on one side = 350/5 + 1 = 70 + 1 = 71
Total number of palm trees planted = 71 x 2 = 142
We add 1 because there is a tree at the beginning and at the end of the street also
Think of it like this, if the street was 5 meters long, how many trees would it have on one side? 5/5 + 1 = 2

Solution to Question 28

Each day the snail walks up 2.5 meters but at night it slips down 1.5 meters.
Net distance traveled each day = 2.5 – 1.5 = 1m
In 4 days the snail will climb 4 meters.
On the 5th day, the snail will climb 4 + 2.5 = 6 meters and will be out of the well and will not slide back inside.
Therefore the snail will come out of the well on the 5th day.
The way to solve such questions is to look at the total distance – the distance the snail travels up each day and divide this by the net distance travelled + 1 more day
Total distance – the distance travelled up in one day = 6 – 2.5 = 3.5 meters.
Now the net distance travelled by the snail per day is 1 meters. So it will travel this distance in 3.5 or 4 days. And on the 5th day, it will go out of the well.

Solution to Question 29

As you observe the pattern,
1st figure has 1 square
2nd figure has 3 squares
3rd figure has 5 squares, so on
The pattern they follow is that the number of squares is one less than double the number of the figure.
so the nth figure will have 2n – 1 squares.
i.e. double the no – 1
The 50th figure will have = 2 x 50 – 1 = 100 – 1 = 99 squares

Solution to Question 30

On her way to school Sandra marked the first tree, and then every second tree
So the trees with mark on them are:
1,3,5,7,9,11,13,15,17,19
On her way back, again, she marked the first tree, and then every third tree.
So the trees with mark on them are:
19,16,13,10,7,4,1
From 1 to 19, the numbers that do not appear in either of the list are
2,6,8,12,14,18
Number of trees that have no mark on them are 6.

Solution to Question 31

The distance between any two consecutive posts is 6 meters
Number of gaps between 1st to 10th lamp post = 10 – 1 = 9
The distance between the 1st and the 10th lamp post = 9 x 6 = 54 meters

Solution to Question 32

The distance from A to B is 15 cm.
The distance from C to A is 8cm.
The distance from B to C = 15 – 8 = 7 cm

Solution to Question 33

3 girls have an average height of 150 centimeters.
The sum of heights of all girls = 150 x 3 = 450 centimeters
4 boys have an average height of 170 centimeters
The sum of heights of all the boys = 170 x 4 = 680 centimeters
The average height of all the girls and boys put together
= (680 + 450)/7
= 1130/7
= 161.42 centimeters

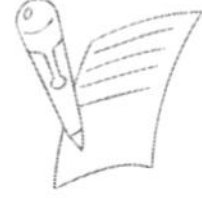

Solution to Question 34

Sandra took the elevator to the 8th floor.
Down 5 floors takes her to = 8 – 5 = 3rd floor
Up 10 floors takes her to = 10 + 3 = 13th floor
And finally down 1 floor takes her to = 13 – 1 = 12th floor
If 12th floor is the middle of the building then the number of floors = 11 + 1 + 11 = 23

Solution to Question 35

Bernard, Suzy and Amie will both be at the lesson on the same day when that day is the LCM of 9, 12, 15

9 = 3 x 3

12 = 2 x 2 x 3

15 = 3 x 5

LCM = 3 x 3 x 2 x 2 x 5 = 180

So every 180 days, the three of them will go for lesson on the same day.

The least number of days before all three will have their music lessons on the same day = 180 days

Solution to Question 36

The average temperature from Monday to Wednesday was 34°

Sum of temperature from Monday to Wednesday = 34 x 3 = 102°

The average temperature from Thursday to Saturday was 37°.

Sum of temperatures from Thu to Sat = 37 x 3 = 111°

The average of all 7 days Monday to Sunday was 35°

The sum of temperatures on all seven days = 35 x 7 = 245°

The temperature on Sunday =

Sum of temperature on all seven days – the sum of temperature on Mon to Wed – the sum of temperature from Thu to Sat

= 245 – 102 – 111 = 32°

The temperature on Sunday was 32°

Solution to Question 37

In three math exams, Tim scores 84, 92, and 94.

If the average of the scores in 4 tests is 91, then the total score in the 4 tests = 91 x 4 = 364

The score he needs on 4th exam to get an overall average of 91 =

364 – 84 – 92 – 94

= 364 – 270

= 94

Tim will need to score 94 in the fourth exam to get an average score of 91 for all the four exams = 94

Solution to Question 38

A box of pencils can be divided equally among 4, 5, 6 or 7 children with no pencils left.

This means the number of pencils is a multiple of 4, 5, 6 and also of 7.

So the number of pencils will be the LCM of 4,5,6,7

4 = 2 x 2

6 = 2 x 3

5 = 5 x 1

7 = 7 x 1

LCM = 2 x 2 x 3 x 5 x 7 = 420

The least number of pencils that the box could have = 420

Solution to Question 39

The number of books leave a remainder of 1 when divided by 3, 4, 5 and 7.

So the number of books will be the LCM of 3, 4, 5 and 7 + 1

LCM of 4,5,6,7

4 = 2 x 2

5 = 5 x 1

6 = 2 x 3

7 = 7 x 1

LCM = 2 x 2 x 3 x 5 x 7 = 420

Number of books George had = 420 + 1 = 421

Solution to Question 40

If there are 15 children, each child gets 20 chocolates.

Total number of chocolates = 15 x 20 = 300

If there are 10 children.

Number of chocolates each child will get = 300/10 = 30

Solution to Question 41

Rebecca was 128 centimeters last year and in the last year she has grown by 7 centimeters

Rebecca's current height = 128 + 7 = 135 centimeters

Kate is 14 centimeters shorter than Rebecca

Kate = 135 – 14 = 121 centimeters

James is 11 centimeters taller than Kate

James = 121 + 11 = 132 centimeters

John is 15 centimeters shorter than James = 132 – 15 = 117 centimeters

John is 117 centimeters tall

Solution to Question 42

The LCM of 6, 9, 15, 20 is

6 = 2 x 3

9 = 3 x 3

15 = 3 x 5

20 = 2 x 2 x 5

LCM = 2 x 3 x 3 x 2 x 5 = 180 minutes or 3 hours

The time at which the bells will ring again at the same time = 6:00 + 3 = 9:00 AM

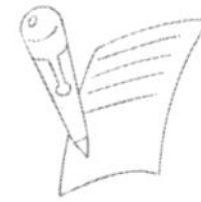

Solution to Question 43

See the below for solution.

Vivian	1st March
Shawn	20th March
Felicia	20th July
Ella	17th May

Ella are born on May 17th.

Solution to Question 44

Since 2 students have both a tape recorder and a pocket calculator, and 7 have a tape recorders, the number of students who have only tape recorder = 7 – 2 = 5. The number of students who have a pocket calculator = 15 but there are 2 who have both a tape recorder and a pocket calculator.
The number of students who have only a pocket calculator = 15 – 2 = 13
The number of students who have a pocket calculator or a tape recorder or both = 13 + 5 + 2 = 20
The class has 30 students. So the number of students who have neither = 30 – 20 = 10

Solution to Question 45

Let W be the weight of the empty box and A be the weight of the apples
W + A = 50
When half the apples are taken out, the weight of apples left will be A/2
W + A/2 = 35
2W + A = 70
W + A = 50
Subtracting the second equation from the first we get
W = 70 – 50 = 20
Weight of the empty box = 50 Kg

Solution to Question 46

Rachel scored 97 marks for Science and 86 marks for English.
Total marks = 356
Let the marks for Math be M and for Social Studies be S
97 + 86 + M + S = 356
M + S= 356 – 183 = 173
She scored 25 more marks for Math than for Social Studies
S = M – 25
Put this value of S in the equation M + S = 173
M + M – 25 = 173
2M = 173 + 25 = 198
M = 198 / 2 = 99
Rachel scored 99 marks in Math.

Solution to Question 47

Number of students who participated only in orchestra = 15 – 6 = 9
Number of students who participated only in sports = 25 – 6 = 19
Number of students who participated in at least one event = 9 + 6 + 19 = 34

Solution to Question 48

If the page number of the left side is n, then the page number on the right side will be (n+1)
We have to find 2 consecutive numbers whose product is 420.
Factors of 420 = 2 * 2 * 5 * 3 * 7
The two numbers are 20 and 21.
The number of the page on the left side is 20.

2. Algebra

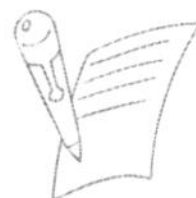

Solution to Question 49

If twice as many coins landed heads up as landed tails up;
then one third of the coins landed tails up and two-thirds as heads up.
Number of coins that landed heads up = 2/3 x 27 = 18

Solution to Question 50

The only animals on farmer Jim's farm are cows and chicken.
Jim counts a total of 55 heads and a total of 136 feet.
Let the no of cows be X.
Number of chickens = 55 – X
A cow has 4 legs; chicken has 2 legs
4X + 2(55 – x) = 136
4X + 110 – 2X = 136
2X = 136 – 110
X = 26/2 = 13
The number of cows on Justin's farm = 13

Solution to Question 51

Jennifer is buying juice cans for her son's birthday party.
Juice cans are sold by the carton.
Let the no of cans in the carton be X.
If she buys 3 cartons of juice cans, she will be short of 3 cans, therefore we add 3 more = $3X + 3$
If she buys 4 cartons, she will have 3 cans extra, so we deduct 3 = $4X - 3$
$3X + 3 = 4X - 3$
$4X - 3X = 3 + 3$
$X = 6$
Number of cans in each carton = 6

Solution to Question 52

Let Set (M) be students taking Mandarin.
Set (F) are those taking French
Set $(M \cap F)$ = Set (M) + Set (F) – Set $(M \cup F)$
$= 9 + 19 - 24 = 4$
Number of Mrs. Smith's students who are taking both Mandarin and French = 4

Solution to Question 53

There are twice as many boys in a room as girls.
Let the no of girls be X.
Number of boys = $2X$
If 5 boys leave the room there would be an equal number of boys and girls in the room.
$2X - 5 = X$
$2X - X = 5$
$X = 5$
Number of boys in the room at first = $5 \times 2 = 10$

Solution to Question 54

Henry had twice as many marbles as Jason.
Let the number of marbles with Jason be 'X'
Henry = 2X
Number of marbles with Kenneth had 3 times as many marbles as Jason.
Kenneth = 3X
They had 846 marbles altogether
$X + 2X + 3X = 846$
$6X = 846$
$X = 846/6 = 141$
Number of marbles with Kenneth = $3 \times 141 = 432$

Solution to Question 55

Jessica arranges stamps into 5 albums with 15 stamps in each album
Number of stamps arranged in album = $5 * 15 = 75$
She has 25 stamps left
Total number of stamps = $75 + 25 = 100$
She then arranges them into equal number of albums with 5 stamps and 15 stamps in each album.
Let X be the no. of albums
$15X + 5X = 100$
$20X = 100$
$X = 100/20 = 5$
Jessica needs 5 albums in all to put all her stamps.

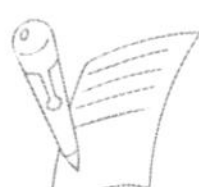

Solution to Question 56

Alison bought 50 potatoes.
She used 2/5 of them to bake = $2/5 * 50 = 20$
Let the number of potatoes she used to bake curry puffs = X
She had 12 potatoes left.
$20 + X + 12 = 50$
$X = 50 - 32$
$X = 18$
Number of potatoes she used for curry puffs = 18

Solution to Question 57

Let the number of balloons that Alice and Susan have each be X.

After Alice used 16 balloons, Susan had 3 times as many balloons as Alice.

$3(X - 16) = X$

$3X - 48 = X$

$2X = 48$

$X = 24$

The number of balloons Susan had in the beginning = 24

Solution to Question 58

Let the weight of the small ball be X.

Weight of the big ball = 2X

$3 * 2X + 4X = 80$

$6X + 4X = 80$

$10X = 80$

$X = 80/10$

X = 8kg

The weight of the big ball = 8 * 2 = 16kg

Solution to Question 59

Let the equal no of stamps bought by Ben and Samuel be X.

$150 + X = 2(50 + X)$

$150 + X = 100 + 2X$

$150 - 100 = 2X - X$

$X = 50$

Each of them bought 50 stamps

Solution to Question 60

One small bottle holds 1.5 liters water.
Let the large bottle hold X liters.
One large bottle and two small bottles can hold the same amount of water as 6 small bottles.
X + 2 x 1.5 = 6 x 1.5
X = 9 – 3
X = 6 Liters
One large bottle holds 6 liters water.

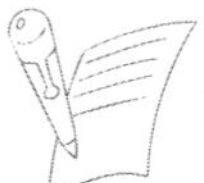

Solution to Question 61

Let the number of red candies be X.
In the jar there were 6 more green candies than red = X + 6
There were one-third as many red candies as yellow candies = X/3
Sarah ate up all the 18 yellow candies first.
X/3 = 18
X = 18 * 3 = 54
Number of green candies still in the jar = 54 + 6 = 60

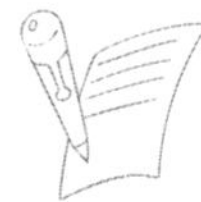

Solution to Question 62

Let the no of pencils with Jennifer be X.
The number of pencils = X – 33
X + X – 33 = 89
2X = 89 + 33
X = 122/2 = 61
Number of pencils with Jennifer = 61
Number of erasers = 61 – 33 = 28

Solution to Question 63

There were 5 times as many strawberries as blue berries in a container.
Let the number of blue berries be 'X'
Number of Straw berries = 5X
Brandon ate half the number of strawberries and 20 strawberries were left in the container.
$5X/2 = 20$,
$5X = 40$
Number of strawberries in the container at first = 40
$X = 8$
Number of blue berries in the container at first = 8

Solution to Question 64

Let the number of coins Lisa has be 'X'
Number of coins with Jessie = 2X
Number of coins with James =2X + 20
$X + 2X + 2X + 20 = 50$
$5X = 50 - 20$
$X = 30/5 = 6$
No of coins Lisa has = 6

Solution to Question 65

William spent 4 times as much on a magazine than on a towel.
Let the amount of money spent on towel be 'X'
Amount of money spent on the magazine = 4X
William had twice as much left as the amount of money he spent. He had $90 left.
Money spent = 90/2 = $45
X + 4X = $45
5X = $45
X = 45/5 = $9
The magazine cost = 4 x 9 = $36

Solution to Question 66

There were 720 orange cakes and twice as many yellow cakes.
Number of Yellow cakes = 720 x 2 = 1440
The rest were green cakes.
Number of green cakes = 2500 – 1440 – 720
= 340

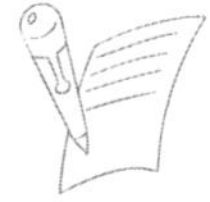

Solution to Question 67

Let the salary be 'X'.
Janice spent half of her salary on food and transport = X/2
She gave the remaining amount to her brother = $450
Savings in the bank = $380.
X/2 + 450 + 380 = X
X – X/2 = 830
X/2 = 830
X = 2 x 830
X = $1660
Janice's salary = $1660

Solution to Question 68

Number of children = 3560 – 1870 = 1690
As, there were 490 more boys than girls
Let the number of girls be 'X'
Number of boys = X + 490
X + X + 490 = 1690
2X = 1690 – 490
2X = 1200
X = 600
Number of girls on the trip = 600

Solution to Question 69

Anna had bought 8 packets of 4 pencils each
Total number of pencils = 8 x 4 = 32
Anna gave 5 pencils to each of her friends
Let the number of friends be 'X'
Number of pencils given = 5X.
Anna then had 7 pencils left.
32 – 5X = 7
5X = 32 – 7 = 25
X = 25/5 = 5
Number of friends she gave the pencils to = 5

Solution to Question 70

Total number of children Initially = 35 x 7 = 245
Total number of children thereafter = 245 + 4 + 16 = 265
Number of children altogether = 265

3. Fractions, Decimal and Percentages

Solution to Question 71

A school drama club has 150 members.
Number of boys = (60/100) x 150 = 90
Number of girls = 150 – 90 = 60
Difference in the number of boys and girls = 90 – 60 = 30
There are 30 more boys than girls

Solution to Question 72

Let the numnber of sweets with Gavin be X.
Gavin gave half of his sweets to Tim = X/2
Tim gave half of his sweets to Emma = 1/2 x X/2 = X/4
Emma gave 1/4 of the sweets to Lily and kept the remaining 12 = 1/4 x X/4
Sweet remaining = 3/4 x X/4 = 3X/16
3X/16 = 12
X = 12 x 16/3 = 64
Number of sweets Gavin had in the beginning = 64

Solution to Question 73

25% is equals to 1/4.
In order to choose the class prefect for grade 5 at least 25% of the total students must be present to vote.
There are 18 students in Grade 5.
Number of students that must be present to elect the class prefect = ¼ x 18
= 4.5 = 5
As there can be no half student, we round of the figure to 5.
Therefore, 5 students must be present to elect the class prefect.

Solution to Question 74

A movie theatre is full.
Let the capacity of theatre be X.
46 girls occupy 2/5 of the seats.
2X/5 = 46
X = 46 x 5/2
X = 115
The capacity of the movie theatre = 115

Solution to Question 75

Jake plays Tennis for 5 hours.
Katie plays tennis twice as long as Jake = 5 x 2 = 10 hours
Ben plays only two-fifth of the time that Katie plays = 2/5 x 10 = 4 hours
Tim plays tennis the quarter of the time that Ben plays = 4 x 1/4 = 1 hour
Tim plays for 1 hour

Solution to Question 76

Christian added 3grams sugar to 15 liters water.
3g = 0.003 kg
As 1 liter of water weighs 1 kg,
15 L = 15 kg of water,
By comparing the masses of sugar and water in the solution, we can obtain the percentage of sugar in the solution.
The percentage of sugar in the solution obtained = 0.003/15 * 100 = 0.02%

Solution to Question 77

Daniel has 200 books in his house library.
70 of them are fiction books.
Non fiction books = 200 – 70 = 130
% of Daniel's books that are non-fiction books = 130/200 x 100 = 65 %

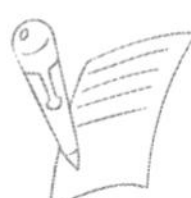

Solution to Question 78

A 10 x 10 = 100 squares
A 10 by 10 grid has 42% of the squares shaded.
42/100 x 100 = 42
Number of the squares shaded are 42.
Number of squares not shaded = 100 -42 = 58

Solution to Question 79

Total number of votes = 100%
Lily received 40% of the votes
Felicia received 35% of the votes
Claire received the rest of the votes.
Percentage of votes Claire received = 100 – 45 – 35 = 20 %

Solution to Question 80

Percentage of votes different flavours got in the survey.
Chocolate chip = 22%
Strawberry = 1/4 x 100 = 25%
Vanilla = 37/100 = 37%
Order of the flavors from most to least popular
= Vanilla, Strawberry, Chocolate Chip

Solution to Question 81

Mrs. Franklin baked a total of 1500 apple tarts.
She gave 45% of them to her relatives = 45/100 x 1500 = 675
Number of tarts remaining = 1500 – 675 = 825
She sold 60% of the remaining tarts = 60/100 x 825 = 495
She gave 90 tarts to her neighbor.
Number of tarts she had left = 1500 – 675 – 495 – 90 = 240

Solution to Question 82

Some students were surveyed on their favorite subject.
Two fifth students preferred Math = 2/5 x 100 = 40%
0.45 preferred English = 45/100 = 45%
48% preferred science.
Order of the subjects from least favorite to most favorite
= Science, English, Math

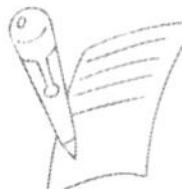

Solution to Question 83

Julian got the scores of his final exam.
The average of his Science, English, and History scores is 65%.
(S + E + H)/3 = 65
S + E + H = 195
When his math, Science, English, and History grades are combined, the average score is 66%.
(M + S + E + H)/4 = 66
M + S + E + H = 264
Julian's Math score = 264 – 195 = 69%

Solution to Question 84

A school bus goes around dropping children at their bus stops.

Let the total no of students be X

At the first stop, 1/4 of the students got off = X/4

Remaining = X – X/4 = 3X/4

At the second stop, 2/3 of the remaining students got off = 2/3 * 3X/4 = X/2

Remaining = 1/3 x 3X/4 = X/4

At the third stop, 1/2 of the remaining students got off = 1/2 x X/4

At the fourth stop, all the remaining students got off = 1/2 x X/4

To have the least number we need to equate the above equation to 1

X/8 = 1

X = 8

The least number of students that could have been on the bus = 8

Solution to Question 85

At 8:30 AM, 200 students started running for a charity run.

By 9:30 AM, the number of students increased by 30%

= 200 + 200 x 30/100

= 200 + 60 = 260

At 10:30 AM, 15% more students joined

= 260 + 260 x 15/100

= 260 + 39 = 299

After another hour, 25% more students joined.

= 299 + 299 x 25/100

= 299 + 75 = 374

Number of students who ran for the charity run = 374

Solution to Question 86

Let Richard's salary be $X.
Amount spent on food = X/5
Amount money spent on clothes = 2X/3
Total spent = 2X/3 + X/5
= (10X + 3X)/15
=13X/15
Remaining amount = X – 13X/15 = 2X/15
Half of the remaining amount was given to his wife = 2X/15 * 1/2 = X/15
The rest was saved.
Amount saved = $ X/15 = $520
X/15 = 520
X = 520 * 15 = $7800
Richard's salary wa $7800.

Solution to Question 87

Let the total no of eggs Samuel had in the beginning be X.
Samuel sold 1/4 of his eggs on Monday = X/4
Remaining number of eggs = 3X/4
1/4 of the remainder on Tuesday = 1/4 * 3X/4
The rest were sold on Wednesday = 3/4 * 3X/4 = 9X/16
He sold 144 eggs on Wednesday
9X/16 = 144
X = 144 * 16/9 = 256
Number of eggs he had in the beginning = 256

Solution to Question 88

Anna had read ¾ of the storybook.
Fraction of the book she is yet to read = ¼
Since she had 135 pages more to read, ¼ of the book = 135.
So the number of pages in the book = 135 x 4 = 540.
The number of pages she has read = 540 – 135 = 405

Solution to Question 89

The length of 2 parts of the rope is 2/9 x 3 = 6 meters

Solution to Question 90

Let the number be X.
2/3 x X = 1/2 x X + 12
4X = 3X + 72
X = 72
Third multiple of the number = 72 x 3 = 216
The difference of the number and its 3rd multiple = 216 – 72 = 144

Solution to Question 91

There were a total of 2250 seats in a theatre.
10% of the seats were first class = 10/100 x 2250 = 225
30% of the seats were second class = 30/100 x 2250 = 675
and the rest of the seats were third class = 2250 – 225 – 675 = 1350
2 years later, 100 first class seats and 125 second class seats were added.
First class seats in the end = 225 + 100 = 325
Second class seats in the end = 675 + 125 = 800
The total number of seats in theatre = 2250 + 125 + 100 = 2475
Percentage of seats that were third class in the end = 1350/2475 x 100 = 54%

Solution to Question 92

Let Mrs Smith's savings be X.
Mrs. Smith spent 2/3 of her savings to buy furniture for her house. = 2X/3
Remaining savings = X – 2X/3 = X/3
She then spent 1/2 of her remaining savings to get it painted = 1/2 x X/3
The cost of painting the house was $350
X/6 = 350
X = 350 x 6 = $2100
Mrs Smith's original savings = $2100

4. Ratio and Proportion

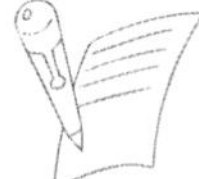

Solution to Question 93

Time taken to make 525 cups = 525 / 150 x 7 = 24.5 minutes

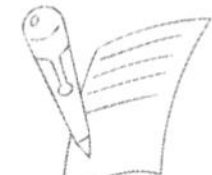

Solution to Question 94

Ratio of brown dogs to white dogs to grey dogs is as follows,

B : W : G
2 : 1 : 3

There is a total of 6 units of dogs.
1 unit = 114/ 6 = 19 As grey dogs are 3 units,
Number of grey dogs = 19 x 3 = 57

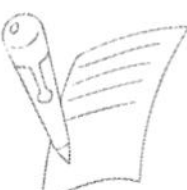

Solution to Question 95

Mr. Fred is buying chocolates for 4 dozen students of his class.
Total number of Students = 4 x 12 = 48
The chocolates come in a packaged of 10 in a box.
Total numbber of chocolates required = 50
Least number of boxes he can buy so that each student gets at least 1 chocolate = 50/10 = 5

Solution to Question 96

Before:
Let the number of France stamps with Derrick be 'X'
Canada stamps = 3X
X + 3X = 168
4X = 168
X = 168/4 = 42 France stamps and 3 x 42 = 126 Canada Stamps
After:
Derrick's aunt gave him another 54 Canada stamps.
Number of Canada stamps = 126 + 54 = 180
Canada stamps Derrick had in the end = 180

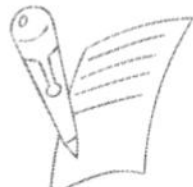

Solution to Question 97

It takes 2000 bees one year to make 6 jars of honey.

Bees	year	jars
2000	1	6
4000	?	60

As the number of bees increases, the rate increases, so it is in direct proportion.
2000 bees make 6 jars in 1 year, therefore 4000 bees make
= 6 x 2 = 12 jars in 1 year
Number of years it will take 4000 bees to make 60 jars of honey = 60/12 = 5 years

Solution to Question 98

Some girls went for rock climbing.
The troop leader asked 7 girls when they wanted to go rock climbing.
Three of the 7 girls wanted to go rock climbing in the morning.
There were 28 girls on the trip.
Number of girls who wanted to go rock climbing in the morning = 3 x 28/7
= 3 x 4
= 12

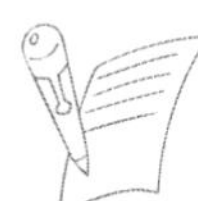

Solution to Question 99

Liz went for fishing with her father at 7.00am.
There is a limit of catching 15 fish per hour at the fishing ground.
Liz needs 110 fish
1 hr ——— 15
? —— 110
Number of hours required = 110/15 = 7 hours
Time by which she will be able to catch her 110 fish = 7 + 7 = 14 hr i.e. 2 : 20pm

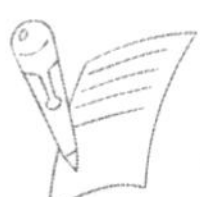

Solution to Question 100

Linda has a dollar's worth of 20 cents and a dollar's worth of 10 cents.
$1 = 5 twenty cent coins
$1 = 10 ten cent coins
The ratio of 20 cents to 10 cents = 5 : 10 = 1 : 2

Solution to Question 101

A novel with 320 pages has about 600 words per page.
Total number of words = 320 x 600 = 192000
Liz read the book at a rate of 300 words a minute.
1 min —— 300
? —— 192000
= 192000/300 = 640 min
1 hr = 60 min
? —— 640
= 640/60 = 10 hrs
Number of hours taken to finish the book = 10 hrs

Solution to Question 102

Let the no of cakes be X.
Laura sold twice as many chocolate cookies as cakes = 2X
three times as many lemon tarts as cakes = 3X
Laura sold 26 lemon tarts
3X = 27
X = 27/3 = 9
Number of cookies she sold = 2 x 9 = 18

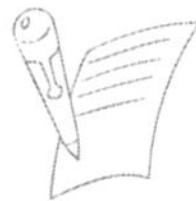

Solution to Question 103

The sum of three numbers is 81 and their ratio is 3 : 7 : 17.
Ratio = 3 + 7 + 17 = 27
The value of the smallest number = 3/27 x 81 = 9

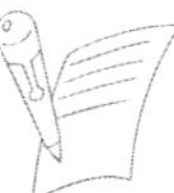

Solution to Question 104

Martha practiced playing guitar on 12 different days in January.
The ratio of days that she practiced the guitar to all the days in January = 12 : 31

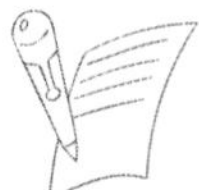

Solution to Question 105

Jimmy can eat one sixth of a hot dog in two minutes.
1/6 ——— 2 min
1 ——— ?
Therefore Jimmy takes 2 x 6 = 12 min to complete the hot dog.
It takes 3 minutes for Lucas to eat one quarter of the hot dog.
1/4 ——— 3 min
1 ——— ?
Therefore Lucas takes 3 x 4 = 12 min to finish the hot dog.
If Jimmy and Lucas start eating one hot dog each, both will finish at the same time.

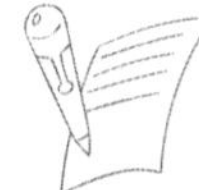

Solution to Question 106

Kathy is 25 years old.
Her sister is 4/5 of her age = 4/5 x 25 = 20 years
The ratio of her sister's age to their mother's age is 4:10
Let her mother's age be X

$$\frac{20}{X} = \frac{4}{10}$$

X = 20 x 10/4 = 50 years
Kathy's mother is 25 years older than her.

Solution to Question 107

Jenny performs ballet 2 days out of every 7 days.
1 month = 4 weeks
Number of days on which Jenny perform ballet = 2 x 4 = 8 days
Number of days on which Jenny does not perform ballet = 28 – 8 = 20
The ratio of days she performs ballet to days she does not perform in a month
= 8 : 20
= 2 : 5

Solution to Question 108

Physics tells us that weight of an object on the moon is proportional to their weights on Earth.
An 80 kg man weighs 30 kg on the moon.
80 ——— 30
60 ——— ?
Weight of a 60 kg man on the moon = 60 x 30/80 = 22.5 kg

Solution to Question 109

Brian types 1 page each minute.
Wendy types 3 pages each minute.
Difference = 3 – 1 = 2 pages
In 1 min Brian is 2 pages behind Wendy
Brian is 2 pages behind Wendy in 1 min
He will be 100 pages behind Wendy
= 100/2 = 50 min
Number of minutes in which Brian will be 100 pages behind Wendy = 50 minutes

5. Money

Solution to Question 110

Mr. Ted bought a computer that was on sale for $910.
The regular price of the computer was $1,495.
Savings = 1495 – 910 = $ 585
Amount of money Mr. Ted saved by buying the computer on sale = $585

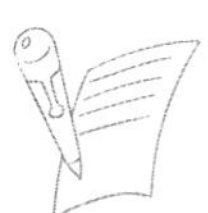

Solution to Question 111

Pete paid for 4 identical pizzas with a $50 bill.
Pete got $3.60 in change
Cost of 4 pizzas = 50 – 3.6 = $46.4
Cost of one pizza cost = 46.4/4 = $11.6

Solution to Question 112

Sam had \$45.
He spent all his money on a photo frame and a book.
The book cost \$3 less than the photo frame.
Let the cost of photo frame be 'X'
Cost of book = X – 3
X + X – 3 = 45
2X = 45 + 3
X = 48/2 = \$24
a) Cost of the book = 24 – 3 = \$21
b) Cost of the photo frame = \$24

Solution to Question 113

Kenneth had \$150 more than Jason and \$70 less than Daniel.
Let the amount of money with Jason be 'X'
Amount of money with Kenneth = X + 150
Amount of money with Daniel = X + 150 + 70 = X + 220
The three boys had a total of \$790.
X + X + 150 + X + 220 = 790
3X + 370 = 790
3X = 790 – 370 = 420
X = 420/3 = 140
Amount of money with Daniel = 140 + 220 = \$360

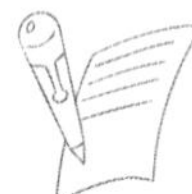

Solution to Question 114

An apple and 2 bananas cost \$2.70. The banana cost 30cents less than the apple.
Let the Apple's cost be ' \$X'
Cost of Banana = \$ (X – 0.30)
\$X + 2 x \$(X – 0.30) = \$2.70
\$X + \$2X – \$0.6 = \$2.70
3X = \$2.70 + \$0.60
3X = \$3.30
X = \$1.10
Cost of an apple = \$1.10

Solution to Question 115

Cost of child's ticket = 15 – 7 = $8

Mr. and Mrs. James took some children to the concert. They paid a total of $86.

Let the number of children be 'X'. Adults = 2

2 x 15 + X x 8 = 86

30 + 8X = 86

8X = 86 – 30

8X = 56

X = 56/8 = 7

Number of children Mr and Mrs James took to the concert = 7

Solution to Question 116

$4.20 + $2.90 + Socks + $9.05 = $18.95

Socks + $16.15 = $18.95

Socks = $18.95 – $16.15 = $2.80

Amount of money Janice paid for the pair of socks = $2.80

Solution to Question 117

Bob's salary in 2010 = 6800 x 12 = $81600

In 2011, he received a pay increase of 18%.

Total salary in 2011 = $81600 x 108/100 = $96288

Amount of more money he earned in 2011 than in 2010 = $96288 – $81600 = $14688

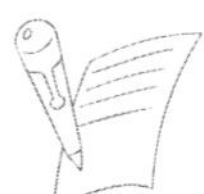

Solution to Question 118

Sales tax charged for the necklace = 220 x (7/100) = $15.40

Solution to Question 119

Sale price of the motor Bike = $500 + 12 x $350 = $4700
As the sales price was 10% of the original price, we will multiply by 10 to find the original price.
Original Price of Motor Bike = $4700 x 10 = $47000
After the sales promotion price of the motor bike= $47000 x (105/100) = $49350

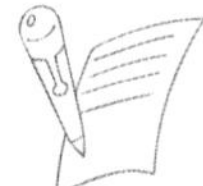

Solution to Question 120

Lucy spent $60 on 120 pencils.
a) Cost of each pencil = 60/120 = $0.50
A pen cost 25cents more than a pencil.
Cost of a pen = $0.50 + $0.25 = $0.75
b) Number of pens she can buy if she spends $60 = 60/0.75 = 80

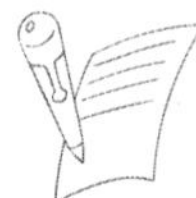

Solution to Question 121

Tim wants to purchase a new Television which costs $3050.
He first needs to pay a down-payment of $1015.
Remaining amount = 3050 – 1015 = $2028
The remaining amount will be paid in monthly installment over 12 months.
Each month installment = 2028/12 = $169

Solution to Question 122

A pair of shoes cost half as much as a pair of trousers.
Let the cost of a pair of shoes be $X
The cost of a pair of trousers = $2X
The total cost of 3 such pairs of shoes and a pair of trousers is $250
3X + 2X = 250
5X = 250
X = 250/5
X = $50
The cost of a pair of shoes = $50

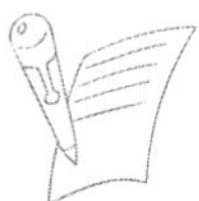

Solution to Question 123

Samuel earns \$2450 a month.
He spends 2/5 of it = 2/5 * 2450 = \$980
He saves the rest = 2450 – 980 = \$1470
Annie earns \$800 less than Samuel = 2450 – 800 = \$1650
Annie spends \$180 more than Samuel each month = 980 + 180 = \$1160
Annie's savings per month = 1650 – 1160 = \$490
Amount of extra money saved more by Samuel in a month than Annie = \$1470 – \$490 = \$980
Amount of money that Samuel saves more than Annie in half a year = \$980 x 6 = \$5880

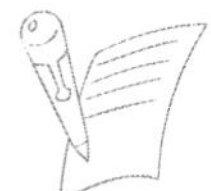

Solution to Question 124

Let the no of buttons Sarah buys be X.
Number of ribbons = 9 – X
Each button costs \$0.60 and each ribbon costs \$0.70 and total money spent = \$5.90
0.6X + 0.7 (9 – X) = 5.9
0.6X + 6.3 – 0.7X = 5.9
0.1X = 6.3 – 5.9
X = 0.4/0.1 = 4
Number of buttons Sarah bought = 4

Solution to Question 125

Cost of a potato = \$P and cost of an orange = \$O
One potato and two oranges cost \$2.30
1P + 2O = 2.3
Two potatoes and an orange cost \$2.50.
2P + 0 = 2.5
Add both the equations:
3P + 3O = \$4.8
Divide both sides by 3
P + O = \$1.60
Cost of a potato and an orange = \$1.60

Solution to Question 126

Let the number of days after which John's saving will be twice of Peters' be X.
Peter has $4 in his piggy bank and John has $2 in his piggy bank.
Peter adds $1 to his piggy bank each day and John adds $3 to his each day.
$4 + 2X = 2 + 3X$
$4 - 2 = 3X - 2X$
$X = 2$ days
After 2 days John's savings will be twice of Peters' savings.

Solution to Question 127

Let the cost of a novel be $N and the cost of a book be $B

$N + 2B = 34$
$2N + B = 86$

$3N + 3B = 120$ (adding both the equations)

Dividing both sides by 3 we get
$N + B = \$40$
The cost of 5 books and 5 novels will be
$5N + 5B = 5 \times 40 = \$200$
Cost of 5 novels and 5 books = $200.

Solution to Question 128

Let the cost of a spoon be X.
One plate costs thrice as much as a spoon = 3X
1 plate and 4 spoons cost $210
$3X + 4X = 210$
$7X = 210$
$X = 210/7 = \$30$
The cost of one spoon = $30

Solution to Question 129

Let the cost of coke be $X.
The ice-cream cone cost $3 more than the coke = X + 3
The total cost of a coke can and an ice-cream cone is $12.
X + X + 3 = 12
2X = 12 – 3
X = 9/2 = $4.5
Cost of a can of Coke = $4.5
Cost of an Ice cream = 4.5 + 3 = $7.5

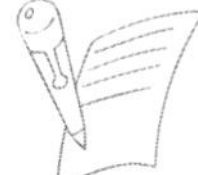

Solution to Question 130

Let the cost of an eraser be X.
The cost of a pencil = X + 1
X + X + 1= 1.1
2X = 1.1 – 1 = 0.1
X = $0.1/2 = $0.05 = 5 cents
Cost of an eraser if 5 cents

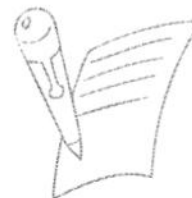

Solution to Question 131

Each notebook costs $0.80 and each pen cost one-fourth the cost of a notebook.
Cost of a pen = $0.8/4 = $0.2
Let the number of pens and bought be X.
Number of notebooks bought is the same.
0.8X + 0.2X = 15
1 X = 15
X = 15
Number of pens George bought = 15

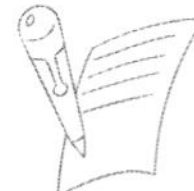

Solution to Question 132

Cost of two T-shirts = $18 x 2 = $36
Total cost of all other shopping items = 31 + 36 + 37 = $104
Total amount of money Jimmy had in his wallet initially = $104 + $32 = $136
Amount of money Jimmy's mother gave him = $136 – $55 = $81

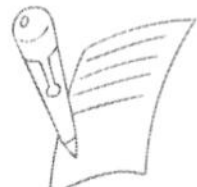

Solution to Question 133

If Ted's savings double every year for 5 years
After, 1st yr = 500 x 2 = $1000
2nd yr = 1000 x 2 = $2000
3rd yr = 2000 x 2 = $4000
4th yr = 4000 x 2 = $8000
5th yr = 8000 x 2 = $16000
Money Ted would save at the end of 5 years = $16000

Solution to Question 134

Cost price of 1 bag = 45/3 = $15
Profit = Selling price – Cost price
= 32 – 15 = $17 for 1 bag
Last month, Mrs Ali Marino sold every purse she bought, and made a profit of $221.
Number of purses she bought = 221/17 = 13

Solution to Question 135

Tim bought 10 apples and his friend Joe bought 7 mangoes.
The price of each apple is $0.55 and the price of each mango is $1.10.
Cost of 10 apples = 10 * 0.55 = $5.50
Cost of 7 mangoes = 1.1 * 7 = $7.70
Total price of the fruits Tim and Joe bought together = $5.50 + $7.70 = $13.20

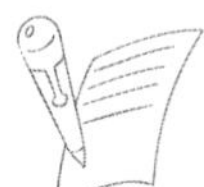

Solution to Question 136

Sarah needs to buy pencils for school.
She can buy 12 pencils for $6
Cost of one pencil = $6/12 = $0.50
Or she can buy 20 pencils for $7.50
Cost of one pencil = $7.5/20 = $0.375 $0.38
Sarah should buy 20 pencils for $7.50

Solution to Question 137

Discounted price of the bicycle = 85 x (75/100) = $63.75
Amount of money Jim received back = $100 – $63.75 = $ 26.25

Solution to Question 138

Red markers cost $1 each, blue marker cost $2 each and golden markers cost $5 each.
If Juliet bought at least one of each
Total cost of 3 marbels = 1 + 2 + 5 = $8
Juliet bought 10 makers. Cost of remaining
7 markers = 18 - 8 = $10
With the above 2 conditions, the possible number of markers Juliet bought are:
3 Blue markers and 4 Red markers.
Number of Red markers Juliet bought = 4 + 1 = 5

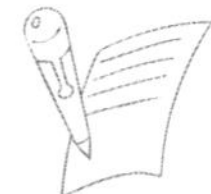

Solution to Question 139

A painter charges $ 335 for materials and $ 25 per hour for his labour.
The total cost of painting a house is $ 415.
Labour cost = 415 – 335 = $80
Number of hours it takes the painter to paint the house = 80/25 = 3.2 hr = 3 hr 12 minutes

6. Measurement and Time

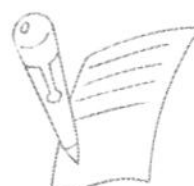

Solution to Question 140

Tia made a sandwich with some cheese and 2 slices of bread that each weighed 2 grams.
Weight of 2 slice of bread = 2 * 2 = 4 grams
She weighed the sandwich and found it weighed 9 grams.
Cheese she used = 9 – 4 = 5 grams

Solution to Question 141

Michael drinks a 350 ml bottle of lemon water every day.

Amount of lemon water he will drink in one week = 350 * 7 = 2450 ml

1 liter = 1000 ml

? Lt = 2450 ml

= 2450/1000 = 2.45 litres

Amount of lemon water he will drink in one week in liters

= 2.45 litres

Solution to Question 142

A car uses 3.5 liters of fuel every 2 kilometers it travels.

3.5 liters ——— 2 km

? liters ——— 75 km

Amount of fuel it uses to travel 75 kilometers = 75 * 3.5/2 = 131.25 liters

Solution to Question 143

Amie has a jug of lemonade.

She does not know how much lemonade she has, but she knows she can fill 12 glasses which have a capacity of 280 ml each.

Amount of lemonade she had = 280 * 12 = 3360 ml i.e. 3.36 liters

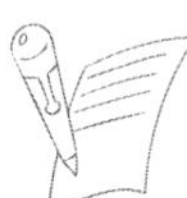

Solution to Question 144

Jug A holds 1800 ml water.

Jug B holds 3/4 more = 1800 + 1800 * 3/4 = 1800 + 1350 = 3150 ml

Amount of water jug B hold = 3150 ml

Amount of water the two jugs hold = 3150 + 1800 = 4950 ml

Solution to Question 145

Sarah creates a Fruit punch.
1 liter = 1000 ml
It contains 1/10 of a liter of apple juice = 1000/10 = 100 ml
2/5 of a liter of orange juice = 2/5 * 1000 = 400 ml
1/8 of a liter of grape juice = 1000/8 = 125 ml
= 100 + 400 + 125 = 625 ml i.e. Jug 3
The jug which is the most suitable for Sarah to serve her fruit punch in is Jug 3.

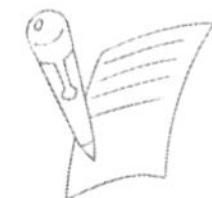

Solution to Question 146

Difference between the fast and the slow watch in 1 hr = 9:02 – 8:59
= 3 minuntes.
1hr —— 3 min
? —— 60 min
60/3 = 20 hrs
After 20 hrs both the watches will have a difference of 1 hr = 8 + 20 = 4:00 a.m. real time
The fast watch will show 4.40am as it gains 20 x 2 = 40 min in 20 hours
The slow watch will show 3.40 a.m.

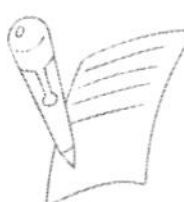

Solution to Question 147

Jerry is 12 years old
Donald is twice as old as Jerry = 2 x 12 = 24 yrs
Mickey is twice as old as Donald = 24 x 2 = 48 yrs
Mickey is 48 yrs old

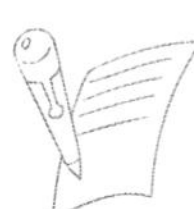

Solution to Question 148

1 hr = 60 min
1 min = 60 seconds
Therefore 1 hr = 60 x60 = 3600 seconds
The correct time 3600 seconds before 1:30 PM = 12:30 pm

Solution to Question 149

A planet far away in the space has 15 months in a year, 7 weeks in a month and 24 days in a week.
Number of days in one year on the planet = 15 x 7 x 24 = 2520
A person on that planet is 7 years old.
Age of the person in days on the other planet = 2520 x 7 = 17640
1 year on Earth = 365 days
Person's age on earth = 17640/365 = 48.32 years

Solution to Question 150

Let the number of years be X after which Susan will be thrice as old as Nicolas.
3 x (4 + X) = 24 + X
12 + 3X = 24 + X
3X – X = 24 – 12
2X = 12
X = 6
After 6 years, Susan will be thrice as old as Nicolas.

Solution to Question 151

You cannot subtract 50 minutes from 35 minutes. Since 1 hour has 60 minutes, convert 6 days 7 hours 35 minutes into
6 days 6 hours 95 minutes (convert 7 hours to 6 hours + 60 minutes)
– 3 days 9 hours 50 minutes

Also, you cannot subtract 9 hours from 6 hours. Since 1 day = 24 hours, so convert 6 days 6 hours 95 minutes into
5 days 30 hours 95 minutes (Convert 6 days into 5 days + 24 hours)

So the problem becomes:
5 days 30 hours 95 minutes
– 3 days 9 hours 50 minutes
= 2 days 21 hours 45 minutes

Solution to Question 152

Stacy and Richard got married on May 18th 2011.
May 18th 2011 is Wednesday
So May 1st 2011 is Sunday
In 2011, February has 29 days
Number of days between 1st May 2011 and 1st May 2012 = 366 days
366 days = 52 weeks + 2 days
So 1st May 2012 is Sunday + 2 days = Tuesday
Mother's Day is on 1st Monday of the month
So Mother's Day 2012 is on 7th May 2012.

7. Speed, Distance and Time

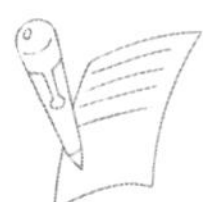

Solution to Question 153

The total distance that the ball had traveled in the air when it hit the ground for the 4th time = 250 + 500 + 500 + 500 = 1750 meters

Solution to Question 154

Machine A can make 2050 ice-creams a day.
This is 110 fewer ice-creams than what machine B can make.
Number of ice-cream machine B can make = 2050 + 110 = 2160
10 ice- creams are packed in a packet and every 1000 packets of ice-creams are placed in a box.
Number of ice creams made per week by:
Machine A = 2050 x 7 = 14350
Machine B = 2160 x 7 = 15120
Total number of ice creams made = 14350 + 15120 = 29470
a) If both machines A and B are used, no of packets of ice-creams after a week = 29470/10 = 2947
b) No of complete boxes of ice-creams after a week = 2947/1000 = 2

Solution to Question 155

Effective rate of making snowballs per hour = 18 – (4 x 3)

= 6 snowballs per hour

Time taken to make 250 snowballs = 250/6 = $41\frac{2}{3}$ hr

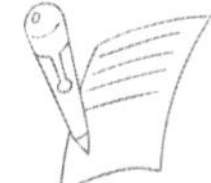

Solution to Question 156

Shelly eats 12 sweets in 4 minutes.
Number of sweets Jane eats in the same time = 4 x 12 = 48

Solution to Question 157

Average speed of Tim for the whole trip = total distance/total time
= (200 + 40 + 360)/(10 + 20 + 10)
= 600/40
= 15 miles/hour

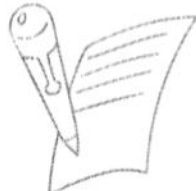

Solution to Question 158

A bus traveling at an average rate of 50 kilometers per hour made the trip to town in 6 hours.
Distance = speed * time
Total distance of the trip = 50 * 6 = 300 kms
If the bus had travelled at 45 kilometers per hour'
time taken for the trip = 300/45 = 6.67 hrs i.e. 6 hr 40 mins
Number of more minutes it would have taken to make the trip
= 6:40 – 6 = 40 mins

Solution to Question 159

A bus and a car leave the same place and travel in opposite directions.
The bus is travelling at 50 km per hour and the car is travelling at 55 kilometers per hour.
Total distance traveled by them in 1 hour = 50 + 55 = 105 km
Remaining journey = 200 – 105 = 95 kms
1 hr ——— 105 km
? Hr ——— 95 km
= 95/105 = 0.9 hour i.e. 54 min
Number of hours in which they will be 200 kilometers apart = 1 hour 54 min

Solution to Question 160

John took a drive to the town at an average rate of 40 kilometers per hour.
Distance = speed * time
Let the distance to the town be 'X'
Time taken to reach the town = X/40
In the evening, John drove back at 30 kilometers per hour.
Time taken = X/30
He spent a total of 6 hours travelling
X/40 + X/30 = 6
3X + 4X = 6 * 120
7X = 720
X = 720/7
X = 102.85 = 103 km
The distance travelled by John = 103 km

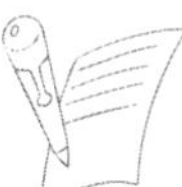

Solution to Question 161

Lily runs at 1m/sec and Trevor runs at 2m/sec
Number of seconds in 2 hours = 2 x 60 x 60 = 7200
Distance Lily runs in 2 hours = 7200 x 1 =7200 = 7.2 km
Distance Trevor runs in 2 hours = 7200 x 2 = 14400 = 14.4 km
The distance they are apart in 2 hours = 7.2 + 14.4 = 21.6 km

Solution to Question 162

The distance between cities A and B is 240 kilometers.
Let the point where both the cars met be at a distance ‘X’ from town A
Distance from town B = $240 - X$
Distance = speed * time
At 11:00am, a yellow car leaves city “A” at a constant rate of 60 km/hr. towards city “B”.
Time taken by the yellow car = $X/60$
At the same time a blue car leaves city “B” toward city “A” at a constant rate of 50 km/hr.
Time taken by the blue car = $(240 - X)/50$
$X/60 = (240 - X)/50$
Cross multiplying we get
$50X = 240 * 60 - 60X$
$60X + 50X = 14400$
$110X = 14400$
$X = 14400/110 = 130.9 = 131$ kms
Time at which the two cars will cross each other = $131/60 = 2.18$ hours i.e. 2 hours 10 minutes

Solution to Question 163

A total distance of 650 kilometers was covered by a plane in 5 hours at two different speeds.
Let the distance traveled by plane for first part of the trip be ‘X’
Distance for the remaining trip = $650 - X$
Distance = speed * time
For the first part of the trip, the average speed was 105 kilometers per hour.
Time taken for the first part of the trip = $X/105$
The remaining trip was flown at an average speed of 110 kilometers per hour.
Time taken for the remaining part = $(650 - X)/110$
$X/105 + (650 - X)/110 = 6$
$22X + 21 * (650 - X) = 2310 * 6$
$22X + 13650 - 21X = 13860$
$X = 13860 - 13650$
$X = 210$ km

Total distance that the plane flew at 105 km/hr speed = 210 km

Total distance that the plane flew at 110 km/hr speed = $650 - 210 = 440$ km

Solution to Question 164

The entire distance was 300 kilometers.
Let the distance travelled from home to the airport be 'X'.
Therefore distance from airport to Paris = 300 - X
Raymond drove from home at an average speed of 30 kilometers per hour to the airport.
Distance = speed * time
Time taken to reach from home to airport = X/30
He boarded an aeroplane that flew to Paris at an average speed of 60 kilometers per hour.
Time taken from Airport to Paris = (300 – X)/60
The entire trip took six hours.
X/30 + (300 – X)/60 = 6
2X + 300 – X = 60 * 6
X = 360 - 300
X = 60 km
Distance from the airport to Paris = 300 – 60 = 240 km

Solution to Question 165

An aeroplane made a trip to Las Vegas and back.
While going to Las Vegas the aeroplane flew at a speed of 430 kilometers per hour.
On the return trip it flew at a speed of 460 kilometers per hour.
The return trip took 6 hours.
Distance = speed * time
Distance = 460 * 6 = 2760 km
Time taken to reach Las Vegas = 2760/430 = 6.41 hours i.e. 6 hours 41 min
Total time taken for the round trip = 6 + 6:41 = 12 hours 42 minutes

Solution to Question 166

The number opposite to the number 6 when folded would be number 3.

8. Area and Perimeter and Volume

Solution to Question 167

Area of the shaded portion = area of triangle + area of rectangle
Area of the triangle = ½ x 4 x 5 = 10 sq. cm
Area of the rectangle = 2 x 1 = 2 sq. cm
Area of the shaded portion = 10 + 2 = 12 sq. cm

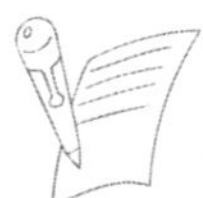

Solution to Question 168

Area of the door = 9 x 4 = 36 sq. feet
Cost per square foot = $15
The cost of a new door
= 36 x 15 = $540

Solution to Question 169

The length of side (AB) = 6 – 1 = 5 cm
The area of the shape ABCDE = area of the complete rectangle - area of the missing square
Area of the rectangle = 7 x 6 = 42 sq. cm
Area of the square = 1 x 1= 1 sq. cm
Area of the shape = 42 - 1 = 41 sq. cm

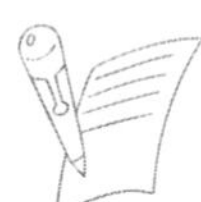

Solution to Question 170

The carpet is rectangular in shape.
The area of the carpet = 25 x 16 = 400 sq. cm

Solution to Question 171

The area of the shape = area of the complete rectangle – area of small the square
Area of the rectangle = 4 x 8 = 32 sq. cm
Area of the small square = 1 x 1 = 1 sq. cm
Area of the shape = 32 – 1 = 31 sq. cm
= 126 sq. cm

Solution to Question 172

Perimeter of the bigger square = 4 x 12 = 48 cm
Perimeter of the smaller square = 4 x 6 = 24 cm
Ratio of the perimeter of the smaller square to the perimeter of the bigger square
= 24 : 48
= 1 : 2
Shaded area = area of the two triangles.
The long edge of the triangle at the bottom = sum of the sides of the two squares
= 12 + 6 = 18 cms
Area of two triangles
= ½ x 18 x 6 + ½ x 12 x 12
= 54 + 72

Solution to Question 173

Area of a triangle = ½ x base x height
Base = 15 – 4 = 11 cms
Height = 6 cms
Area of the shaded portion = ½ x 11 x 6 = 33 sq. cm

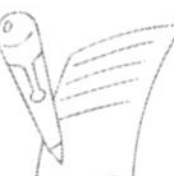

Solution to Question 174

Area of a rectangle = 117 sq. m = length x breadth
Length x 9 = 117
L = 117/9 = 13 m
Perimeter of the rectangle = 2 x (13 + 9) = 2 x 22 = 44 meters

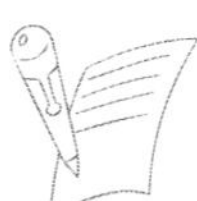

Solution to Question 175

Volume of the block = 30 x 26 x 10 = 7800 cubic cm
Volume of 2 cm cube = 2 x 2 x 2 = 8 cubic cm
Number of 2 cm cube that can be cut from the block = 7800/8 = 975

Solution to Question 176

The sides of a triangle are in the ratio 4:5:6.
Then the sides of the triangle are 4X, 5X and 6X
The shortest side is 6 centimeters
4X = 6
X = 6/4 = 1.5
The remaining sides of the triangle are
5 x 1.5 = 7.5 cm
6 x 1.5 = 9 cm
The perimeter of the triangle in centimeters = 6 + 7.5 + 9 = 22.5 cm

Solution to Question 177

Since the ratio of length to breadth is 1:4, we can take the length and breadth to be X and 4X

Perimeter = 20 cms
2 x (X + 4X) = 20
2 5X = 20
X = 20/10 = 2
Length = 1 x 2 = 2 cm
Breadth = 4 x 2 = 8 cm
The area of the rectangle = 2 x 8 = 16 sq. cm

Solution to Question 178

A farmer has a piece of land that measures 65m by 50m.
Perimeter of the land = 2 x (65 + 50) = 2 x 115 = 230 m
Stakes are put 5m apart all around the land
Number of stakes needed to fence a length of 230 m = 46

Solution to Question 179

Area of the shaded portion = area of square – area of non-shaded triangle
Base of triangle = 12 – 5 = 7 cm
Area of Triangle = ½ x 7 x 12 = 6 x 7 = 42 sq. cm
Area of square = 12 x 12 = 144 sq. cm
Area of the shaded portion = 144 – 42 = 102 sq. cm

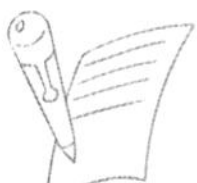

Solution to Question 180

The breadth of the rectangles are in the ratio of 1 : 2
So the value of X will be in the same proportion
X = 7 x 2 = 14 cm

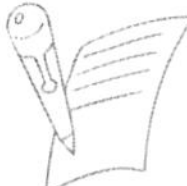

Solution to Question 181

The side of the square is 8 centimeters and the base of the triangle is 5 centimeters.
The area of the non-shaded region = area of square – area of triangle
= 8 x 8 – 1/2 x 5 x 8
= 64 – 20 = 44 sq. cm

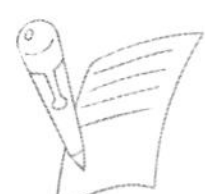

Solution to Question 182

Area occupied by grass = Area of the grass lawn – Area of the pool
Area of the grass lawn = 50 x 20 = 1000 sq. m
Area of the pool = 30 x 10 = 300 sq. m
Area occupied by grass = 1000 – 300 = 700 sq. m

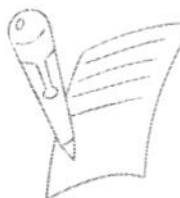

Solution to Question 183

The radius of circle A has the same length as the diameter of circle B.
Let radius of circle B be r, diameter = 2r
Circle A's radius = 2r
The ratio of circle A's area to circle B's area = π x 2r x 2r : π x r x r
= 4 : 1

Solution to Question 184

Let the side of square be S.
Perimeter of the square = 4S = 8
S = 8/4 = 2 cm
Area of the small square = 2 x 2 = 4 cm^2
Area of the big square = 8 x 8 = 64 cm^2
Fewest number of squares, each with a perimeter of 8 centimeters that would completely cover a square with a side of 8 centimeters = 64/4 = 16

Solution to Question 185

The length of a rectangle is 130% of its breadth.
Let the breadth be X.
Length = 1.3X
2 x (1.3 X + X) = 92
2 x 2.3 X = 92
X = 20
Length = 1.3 x 20 = 26 cm
Area of the rectangle = 26 x 20 = 520 cm^2

Solution to Question 186

Area of the remaining cardboard = area of cardboard – area of small squares
Area of the cardboard = 15 x 10 = 150 sq. cm
Area of the small square = 3 x 3 = 9 sq. cm
Total there are 4 squares, therefore area = 4 x 9 = 36 sq. cm
Area of the remaining cardboard = 150 – 36 = 114 sq. cm

Solution to Question 187

The perimeter of square A is 4 times the perimeter of square B.
Let the side of square B be L.
Perimeter of square B = 4L
Perimeter of square A = 4 x 4L = 16L
Side of square A = 16L/4 = 4X
The ratio of the area of square A to the area of square B
= 4L x 4L : L x L
=16 : 1

Solution to Question 188

Sandra's original picture is 4 cm wide and 6 cm long.
The new picture is 10 cm wide
Proportional increase = 10÷4 = 2.5
Length of the new large picture = 2.5 x 6 = 15 cms

Solution to Question 189

If the side of square A is a, its perimeter = 4a.
Perimeter of square B = 2 x 4a = 8a
Side of Square B = 8a ÷4 = 2a

Area of Square A = a2
Area of Square B = (2a)2 = 4a2
Ratio of the area of square A to the area of square B = a2 :4a2
= 1 : 4

Solution to Question 190

Length of the rectangle is 15cm and its breadth is 12cm
Perimeter of the rectangle = 2 * (15 + 12) = 2 * 27 = 54 cm
Area of the rectangle = 12 x 15 = 180 sq cms
Perimeter of a square = Half the area of a rectangle = 180/2 = 90 cms
A wire is used to make 3 squares and 2 rectangles with 20 cm of wire left.
Length of the wire = 3 x 90 + 2 x 54 + 20
= 270 + 108 + 20
= 398 cms

Solution to Question 191

A wire is divided into 2 parts equally.
One of the parts is bent to form two similar squares with sides 10 cm.
Perimeter of one square = 4 x 15 = 60 cms
The length used to make 2 squares = 2 x 60 = 120 cms.
The other part used to make the 3 rectangles will also have the same length. So the perimeter of the three rectangles = 120 cms.

3 * 2 * (L + 5) = 120
6L + 30 = 120
6L = 120 - 30
6L = 90
L = 90/6 = 15 cm

a) The perimeter of each rectangle = 2 * (15 + 5) = 2 * 29 = 40 cm
b) The area of each rectangle = 15 * 5 = 75 sq. cm

Solution to Question 192

The area of the shaded part is 83 sq. cm = area of rectangle – area of square

Let the length of rectangle be L.

L * 9 – 4 * 4 = 83

9L = 83 + 16

L = 99/9 = 11 cm

Area of the whole rectangle = 11 * 9 = 99 sq. cm

Perimeter of the rectangle = 2 * (11 + 9) = 2 * 20 = 40cm

Solution to Question 193

Area of the square = 64 sq. m

Area of the rectangle is thrice as much as that of the square = 64 * 3 = 192 sq. m

From the fig it is clear that the breadth of the rectangle = side of the square

Let the length of square be S

S * S = 8 * 8 = 64

S = 8 cm

Therefore the breadth of rectangle = 8 cm

Let the length of rectangle be L

L * 8 = 192

L = 192/8 = 24 cm

Length of the triangle is 24 cm.

Solution to Question 194

The length of a field is twice its breadth.

Let the breadth be 'B'

Length = 2B

The perimeter of the field is 180 m.

Perimeter = 2 x (L + B) = 180

2 x (2B + B) = 180

3B = 180/2

B = 90/3 = 30 m

The total length of the field = 30 + 30 = 60 meters.

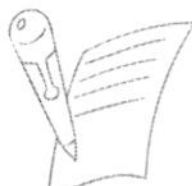

Solution to Question 195

If 's' is the side of a small square, then the length used to make this figure = 12s
12s = 126
s = 126/12 = 10.5

Perimeter of the figure = 8s =
10.5 x 8 = 84 cms

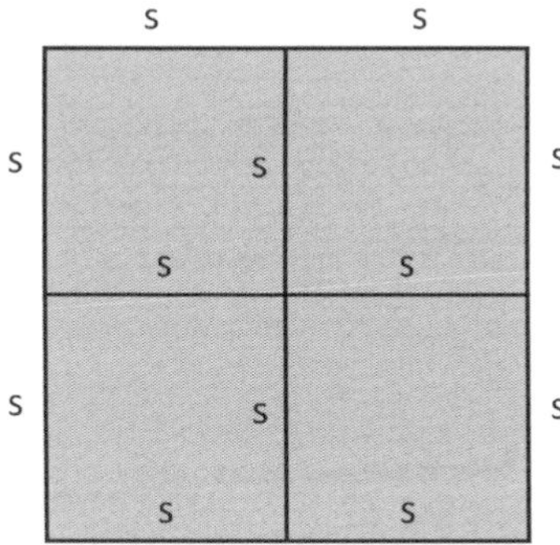

Solution to Question 196

Figure 1: 

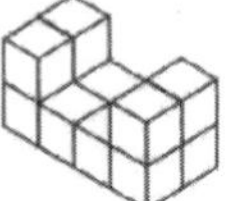

Figure 2:

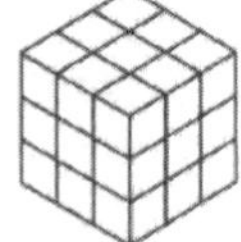

Figure 3:

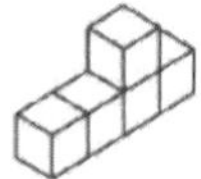

Volume of each small cube = 2 x 2 x2 = 8 units.
The number of cubes to make each figure =
Figure 1 = 12
Figure 2 = 27
Figure 3 = 5
The number of cubes in figure 2 - (number of cubes in figure 1 + number of cubes in figure 3) = 27 - (12 + 5) = 10
What is the volume of figure 1 and figure 3, subtracted from the volume of figure 2 = 10 x 8
= 80 cubic units.

Solution to Question 197

Let the total capacity of tank be X.
3/4 of the tank is filled with water = 3X/4
Volume of the empty tank = X – 3X/4 = X/4
Since 2.8 liters of water is needed to fill this tank to its brim.
X/4 = 2.8
X = 2.8 x 4 = 11.2 liters
The capacity of the tank is 11.2 liters = 11200 cm^3
If the height of the tank is h
Since the area of the base is 800 cm^2,
800 x h = 112000
h = 11200/800 = 14 cm
The height of the tank = 14 cm

Solution to Question 198

Volume of the container = 7 x 10 x 22 = 1540 cubic cm
After some juice was poured into the glasses, the depth of the juice decreased to 15 cm.
Volume of the juice now present in the container = 7 x 10 x 15 = 1050 cubic cm
Volume of the juice poured out = 1540 – 1050 = 490 cubic cm
1 cubic cm = 0.001 Liters
= 490 x 0.001 = 0.49 Liters

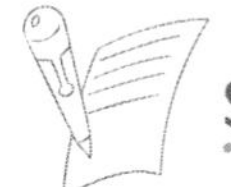

Solution to Question 199

Volume of the tank = 50 x 40 x 25 = 50000 cubic cm
3/4 full = ¾ x 50000 = 37500 cubic cm
Volume of the empty tank = 50000 – 37500 = 12500 cubic cm
Water began to leak from the tank at 400 cm^3 per minute.
Also, it is flowing in at 2.4 liters per minute.
2.4 liters = 2400 cm^3
So the net volume that comes into the tank per minute = 2400 – 400 = 2000 cm^3
Time needed to fill the tank completely
= 12500 / 2000
= 6.25 minutes

Solution to Question 200

Let the breadth of the tank be B.
Volume of water in the tank = 6 X B X 2 = 72
B = 72/12 = 6 cm
Volume of the tank = 6 X 6 X 10 = 360 cm^3
Volume of more water needed to fill up the tank completely = 360 – 72 = 288 cm^3

Printed in Great Britain
by Amazon.co.uk, Ltd.,
Marston Gate.